# Abstract Algebra

# Abstract Algebra

Editor

**Linsen Chou**

# Abstract Algebra

Edited by **Linsen Chou**

Printed in 2017

ISBN: 978-1-68117-180-7

Library of Congress Control Number: 2015949094

© 2016 by
SCITUS Academics LLC,
616 Corporate Way, Suite 2, 4766,
Valley Cottage,NY 10989

www.scitusacademics.com

# Preface

Abstract algebra, a broad division of mathematics, is the study of algebraic structures. Linear algebra, elementary number theory, and discrete mathematics are sometimes considered branches of abstract algebra. Algebraic structures include groups, rings, fields, modules, vector spaces, lattices, and algebra over a field. Algebraic structures, with their associated homomorphisms, form mathematical categories. Category theory is a powerful formalism for analyzing and comparing different algebraic structures.

Universal algebra is a related subject that studies the nature and theories of various types of algebraic structures as a whole. For example, universal algebra studies the overall theory of groups, as distinguished from studying particular groups.

This book, Abstract Algebra, is the set of advanced topics of algebra that deal with abstract algebraic structures rather than the usual number systems. The most important of these structures are groups, rings, and fields. Important branches of abstract algebra are commutative algebra, representation theory, and homological algebra.

# Table of Contents

# The Literature Review of Algebra Learning: Focusing on the Contributions to Students' Difficulties

*Xiong Wang*
Department of Secondary Education,
University of Alberta, Edmonton, Canada

## ABSTRACT

This paper reviews the research literature with respect to the contributions to the students' difficulties in their algebra learning in order to understand the students' difficulties in algebra learning. To start with, 29 articles selected from the database (ERIC) are categorized into a taxonomy which has been generated from the research literature, which falls into five categories including: algebra content, cognitive gap, teaching issues, learning matters, and transition knowledge. The challenges that students confront with under those categories are unpacked in the review process. In addition, the five categories adopted in this paper could serve as a framework of better understanding students' difficulties in their algebra learning. Finally, the research gap from the literature review is discussed.

## INTRODUCTION

Algebra has been recognized as a critical milestone in students' mathematics learning. However, it has been noted that many students created a serious barrier in the algebraic problem solving and formal algebraic system (Kieran, 1992). Therefore, there has been a great attention paid to addressing students' difficulties in algebra learning. This paper is going to review the research literature that bears on the contributions to the students' difficulties in their algebra learning.

In order to evaluate the related literature, 29 articles are selected from my database searching and then categorized into a taxonomy (see Figure 1) including the five categories: algebra content, cognitive gap, teaching issues, learning matters, and transition knowledge. After that, the taxonomy is used to conduct the whole literature review. Within the taxonomy, each category is not independent. For instance, the category of algebra content is the knowledge base for the other four categories. Meanwhile, one of the articles could be coded into more than one category. For instance, the article " Kieran (1992)" is coded into four categories such as algebra content, cognitive gap, teaching issues, and learning matters.

Even though the development of taxonomy is not exhaustive, but it provides a perspective of viewing the contributions to the students' difficulties in algebra learning within the broad ranges, such as mathematics development, school curriculum, teaching practice and students' learning, and so on.

The rest part of this paper focuses on synthesizing and evaluating the existing researches according to the categories demonstrated in the taxonomy. As for the structure of the synthesization and evaluation, discussions will be made 1) on the didactical cut, structural feature of algebra and the characteristics of school algebra in section 2-algbera content; 2) on the cognitive demands from algebra learning, product-process dilemma, and the students' difficulties in operating with the unknown in section 3-cognitive gap; 3) on the discontinuity between primary arithmetic and secondary algebra, scarcity in algebra teaching, and three ideas from China and Singapore in section 4-teaching issues; 4) on the students' difficulties in operational symbols, simplifying expressions, equity, and word problems in section 5-learning matters; and 5) on the needs required by the transition from arithmetic to algebra such as adaptation to a milieu and social interactions and

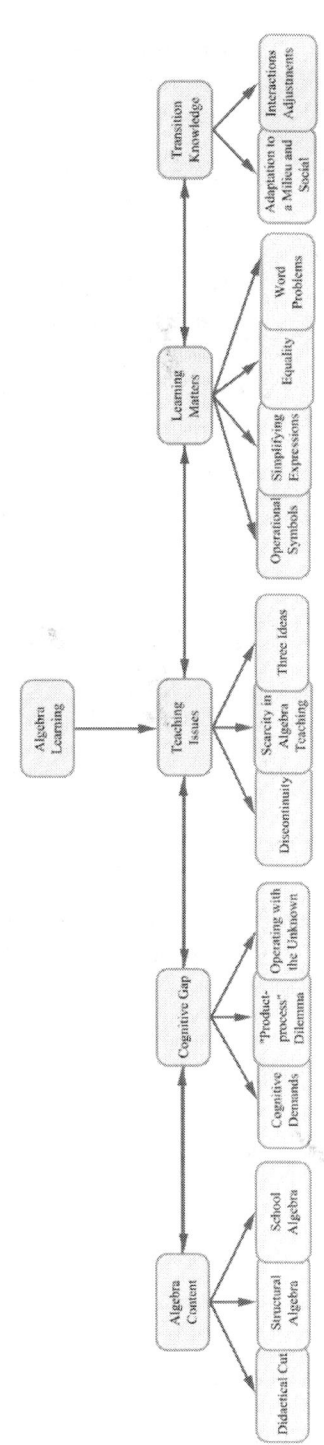

**Figure 1:** Taxonomy of literature review."
v:shapes="_x0000_s1053" class="shape"
v:dpi="96".

adjustments in Section 6-transi- tion knowledge. Finally, the summary and the research gap are provided in Section 7.

## ALGEBRA CONTENT

This category centers on the discussion about the nature of algebra including the didactical cut and structural essence presented in the historical development of algebra. Certainly, school algebra is, to a large extent, influenced by these facts.

### Didactical Cut

Filloy and Rojano (1989) defined one of the fundamental ruptures between arithmetic and algebra is a didacticalcut. The notion referred to the transition that occurred as students face such equations as ax+b=cx+d. Students could successfully solve the equation as ax+b=c using reversal operation as subtracting B from D and dividing by A. This type of equation was called by them as "arithmetical" (p. 19). The reversal operation is not applicable for the non-arithmetical equations as ax+b=cx+d. In order to solve such equations, students have to resort to a truly algebraic idea of operating the unknown (Radford, 2012). Operating the unknown requires students to think analytically, treating the unknown as if it is known (Radford & Puig, 2007). This view provides a specific situation which requires the transition from arithmetic to algebra. Certainly, such requirement stems from the structural nature of algebra.

### Structural Algebra

Kieran (1992) had offered a historical account of the development of algebraic symbolism and its transformational rules, which emphasized the distinguished features of letters between representing unknowns in equation solving and representing givens in expressing general solutions. Furthermore,Kieran (1992) analyzed that the development of algebraic symbolism demonstrated a change from a procedural to a structural perspective on algebra. Meanwhile, the structural development of algebra has a considerable impact on school algebra learning.

## School Algebra

In school mathematics, arithmetic is normally treated as numerical computations (Sfard & Linchevski, 1994). Arithmetic method is used to carry out one or more operations with given numbers to achieve a solution. For elementary algebra, its need is to define the relationships between the unknown and the known data in a problem. As Sadovsky and Sessa (2005: p. 90) pointed out, "the 'object' of arithmetic in primary school is numbers, whereas elementary algebra focuses on relationships between quantities". It is also shown that students' prior exposure in computing binary operations does not prepare them very well to handle algebra (Banerjee & Subramaniam, 2012). For instance, students often apply procedures that have been employed in arithmetic context to simplify algebraic expressions and make the similar mistakes (Fischbein & Barash, 1993).

It is the existing facts in algebra learning that the three categories are the basic barriers and requirements for students' algebra learning. Therefore, those facts will be involved in other sections, which also demonstrate that the category of algebra content is the foundation of the other categories.

## COGNITIVE GAP

Cognitive gap is an obvious obstacle for students' successful transition from arithmetic to algebra as Sfard (1991) suggested. Such cognitive gaps demonstrated from the previous researches are mainly cognitive demands, "product-process" dilemma, and operating with the unknown.

## Cognitive Demands

Learning algebra requires students to take symbolic representations with little or no semantic content as mathematical objects and operate on these objects through processes that usually do not produce numerical solutions (Kieran, 1992). It also requires students to modify their prior experiences in arithmetic context and represent the relationships between quantities in word problems with inverse operations used in arithmetic context (Kieran, 1992). It is clear that the cognitive demands for different operations and representations involved in algebra are intellectual struggles.

## "Product-Process" Dilemma

One cognitive problem was identified by Davis (1975: p. 18) as "name-process" dilemma. The dilemma could be interpreted by the duality "product-process" proposed by Sfard and Linchevski (1993). For instance, an expression such as 8a is both a product for an answer (name) and a process-multiplying 8 by a. Herscovics and Linchevski (1994) showed that, in a teaching experiment, even after instruction, some students could not recognize 8×a as the area (name or product) of a targeted rectangle unless it was embedded in the area formula "S=8×a". Thus, even after an instruction of elementary algebra, students often experience difficulties in operating on a letter representing an unknown in an equation.

## Operating with the Unknown

In order to examine students' experiences in operating the unknown, Linchevski and Herscovics (1996) used equations with only one occurrence of the unknown (e.g. ax+b=c) and equations with two occurrences of the unknown on the same sides (e.g. ax+bx=c) and on different sides of the equal sign (e.g. ax+b=cx+d) to examine the shift in students' procedures. It was found that students could spontaneously group terms that were purely numeric rather than terms in the unknown, which mean "students could not operate spontaneously with or on the unknown" (Linchevski & Herscovics, 1996: p. 41).

In addition, during the process of indicating a clear demarcation between arithmetic and algebra,Herscovics and Linchevski (1994) revealed a difficulty of such pre-algebraic nature as a tendency to detach a numeral from the preceding minus sign in the grouping of numerical terms. For example, in 4+n-2+5=11+3-5, students often add 2 and 5. The high incidence of this mistake demonstrates that the problem is not only common but reflective of unsuspected cognitive obstacles as Herscovics (1989) commented.

It could be drawn to conclusion that the cognitive demands are general requirement from the nature of the algebra. "product-process" dilemma and operating with the unknown stem from the common fact—a letter representing an unknown, which is also part of facts of algebra. Therefore, it could be considered that the cognitive gap is brought about from the nature of algebra.

## TEACHING ISSUE

Limited is the literature on the teaching issues related to the students' difficulties in algebra learning as Kieran (1992) commented. Nevertheless, the limited literatures are sorted into three categories: discontinuity, scarcity in algebra teaching, and three ideas.

### Discontinuity

The traditional arithmetic pays attention to training students' fluency and accuracy in algorithmic computations. While in algebra, students need to have the knowledge and capability of transforming equivalent algebraic expressions. Napaphun (2012) showed that the discontinuity between elementary school arithmetic and the algebra learnt in upper grades was serious. For instance, he mentioned that regarding to the concept of the equal sign, students in most elementary schools were taught to understand the equal sign as a symbol of the calculation. Thus, students habitually thought that an equal sign was always followed by an answer. In fact, recognizing the relation expressed by the equal sign is crucial for algebra learning as Freiman and Lee (2004) suggested.

## Scarcity in Algebra Teaching

Kieran (1992) commented that there was a considerable scarcity not only of teaching models but also of researches on teachers' beliefs and attitudes in algebra. For instance, she mentioned that the teaching models of algebra were not regarded as the way in a different light from ones of arithmetic or geometry. And the teaching of algebra was inclined to focus on such pedagogical issues as the time spent on whole-group instruction or group work, teaching for procedures or understanding, and constructivist or behaviourist approaches to teaching. She further commented that except for the common pedagogical issues, algebra teachers, as other subject teachers, tended to follow a textbook. There are few reports to deal specifically with algebra teaching.

In addition, Rachlin (1989) furthermore pointed out that there was a need for research on algebra curriculum from both content and teachers' perspectives. It is not sufficient to modify the algebra content in the textbook in terms of teachers' heavy reliance on the textbook to change algebra teaching. He suggested that we must understand the nature of teachers' beliefs and attitudes and the roles these beliefs and attitudes play in their teaching. Therefore, exploring the teaching of algebra in some Asian countries such as China and Singapore is a very expedient approach to address teaching strategies in terms of the distinguished performance by the students in those countries in international tests.

## Three Ideas

Recently, Cai and Moyer (2008) commented on the algebra teaching in China and Singapore for purpose of increasing American teachers' knowledge and ability to develop students' algebraic thinking since mathematics achievement in United States was consistently lower than that in those countries. The review suggested that three ideas from Chinese and Singaporean teaching emerging in their reviews were: 1) relating reverse operations to equation solving in Chinese teaching; 2) pictorial equation solving illustrated in Singaporean teaching; and 3) using both arithmetic and algebraic approaches to solve problems in Chinese teaching. Those three ideas were further analyzed by Cai and Moyer (2008) to understand their benefits for students' transition from arithmetic to algebra according to Kieran's (2004) five adjustments for successful transition. It was evidenced that the three ideas had matched with four of the five adjustments. Here, the three ideas are highlighted as a reference for the barren teaching consideration.

The discontinued teaching between arithmetic and algebra is very common in many countries. However, some ideas are verified to be effective in algebraic teaching in China and Singapore, which is the contribution to the scarcity in algebraic teaching.

## LEARNING MATTERS

There is large bulk of researches bearing on students' algebra learning, particularly on students' misconceiving of various concepts in school algebra. It has been typically indicated from the literature we have targeted that students lacked the relevant understanding of operational symbols, simplifying expressions, equality and equation solving. All the evidence are provided from the following discussions.

## Operational Symbols

Booth (1984) reviewed that school algebra was sometimes taken as generalized arithmetic. This meant that the general statements in algebra represented given arithmetical rules and operations. Therefore, students' prior experiences of using symbols in arithmetic would impact on their understanding of the meaning associated with formal symbols in algebra. For example, plus sign is typically interpreted as actions to be operated in arithmetic, which is not used in algebra (Behr et al., 1980). Specifically, Booth (1988) pointed out that, in arithmetic, students were taught to present answers in a single term, such as 3 + 5 was not an acceptable answer. Thus, students were unlikely to recognize a + b to represent a total number of items in two sets owning a and b items, respectively, which further meant that students were unable to regard a + b as a mathematics object in algebra.

In addition, Kiichemann (1981) carried out a large-scale study to examine students' interpretations of literal terms. He found that a great deal of students (13 - 15 years old) could interpret letters as specific unknowns rather than as generalized numbers. His further finding was that majority of students treated letters as concrete objects or overlooked them, which meant that many students were unable to interpret literal expressions as numerical input-output procedures—the first stage in Sfard's (1991) developing process of a structural conception of algebraic expressions.

## Simplifying Expressions

Greeno (1982) conducted a study with beginning algebra students to test their conception of structure of relations in problems. He found that students were short of structural understanding of algebra. For instance, they partitioned algebraic expressions into separately component parts. And more often, students' operation of simplification seemed to be quite at random. For example, 4(6x-3y) +5x was simplified as 4(6x-3y+5x) at one time, and as 4(6x+5x)-3y at another time.

Wenger (1987) also described the students' arbitrary strategies when they dealt with simplifications due to the fact that they could not recognize the right things in algebraic expressions. And students were incapable to transform the simplification knowledge they had learned in one context, as polynomials, to another one, as radicals.

Another typical error in simplifying algebraic expressions was concatenation as Welder (2012) illustrated. For example, 39x-4 was concatenated quite often by students into 35x and 2yz-2y concatenated into z. Carry, Lewis, and Bernard (1980) confirmed that such kind of error occurred not only by beginning algebra students but college students as well. In their study, such error was the most predominant one that students made during their simplifying expressions at different stages of equation solving. Furthermore, they indicated that such error could be caused by students' over generalizing certain validated operations to achieve a generic operation. Thus, the arbitrary strategies and concatenation approaches are the typical behaviours that students demonstrate in the simplification expressions.

## Equality

Equality is one of the requirements for generating and sufficiently interpreting structural representations such as equation (Kieran, 1992). It is normally referred to as the left-right equivalent of the equal sign. However, it is shown from researches that the equal sign is too often misinterpreted by students at all levels of education although high school and college students could be more willing to accept the equal sign as formal symbol for equivalence than younger students as Welder (2012) commented.

Behr, Erlwanger and Nichols (1980) revealed that beginning algebra students took the equal sign as a procedural indicator. For example, students were reluctant to accept expressions such as 3 + 4 = 2 + 5 or 3 = 3. They would like to change equality 3+4=2+5 to be separated into two equalities 3+4=7 and 2+5=7; equality 3+0=3 (Welder, 2012) in terms that they would think that the right side should be the answer.

In addition, Falkner, Levi, and Carpenter (1999) further offered specific data for such limited interpretation of the equal sign. In their investigation, all the participations (145 American students from grade 6) could not correctly fill the number sentence 8+4=_+5. The typical answer for this question was 12 or 17. In addition, the similar situation was presented in the analysis from Li, Ding, Capraro and Capraro (2008). It was evidenced that there were only 25 out of 105 American Grade 6 students could correctly fill the first blank in such number sentence 3+_=4+4=_. However, 91 out of 105 students could give a correct answer 8 for the second blank.

In a word, the misunderstanding and ill operation of equal sign impede students from access to the concept of equity which is the core component of the concept of equation in algebra learning.

## Word Problems

Word problems are regarded as stumbling blocks in algebra to access to higher mathematics, even leading students to drop out of mathematics (Cai et al., 2004). The formal approach used in word problem solving is to formulate an equation or system of equations and operations (Kieran, 1992). However, the students' prior arithmetical experiences posed a great influence on their world problem in secondary.

Khng and Lee (2009) reviewed on the influence of secondary students' prior arithmetical experiences on their word problem solving in Singapore. They commented that algebra word problems were taught in primary school with arithmetic methods, such as counting techniques, guess and test, working backwards, and grouping and model method. And students were very proficient with these methods and regard them as prepotent strategies. Consequently, given the accruing prepotent strategies from primary mathematics, students thought firstly about these strategies rather than algebraic equation formulation when they were presented with word problems in secondary school.

In addition, secondary school teachers often find that beginning algebra students are not motivated to learn the skills needed to solve algebra word problems (Ng & Lee, 2009). This is partly due to the fact that, with the affordance of a concrete and visual representation for the unknowns and arithmetic procedures to solve for the unknowns, students can avoid engaging with the representational and transformational activities, generalising and justifying activities, activities which students find challenging (Kilpatrick, Swafford, & Findell, 2001).

Except for the influence from the arithmetic thinking, students have difficulties in formulating an equation or equations system for a word problem. Reed (1987) found that students had difficulties in recognizing and generating the similar structure among problems with different context. Students often resort to different approaches to access the relations or structures involved in a word problems. For example, syntactic translation and substitution of various numbers are used to verify the adequacy of the equations (Reed, Dempster, & Ettinger, 1985). Tables of relations are also found to be used by students to generate equations for problems; however, repre- senting correct is quite challenging for students (Hoz & Harel, 1989). Moreover, from the cognitive perspective, it is evidenced that students have considerable difficulty in specifying relations among variables (Chaiklin, 1989).

Equation solving is another barrier for students' word problem solving even though they could formulate a correct equation. Students have generally been found to lack the capability to generate and maintain a holistic overview of the structures of an equation, which impacted on the next algebraic transformation to be carried out (Kieran, 1992). For the multi-operation equations, it is noted that students often make very poor strategic decisions in simplifying algebraic expressions and operations (Carry, Lewis, & Bernard, 1980).

Word problem solving is the application field of students' algebraic knowledge including building the relationships between quantities, expressing the relationships by equations, and solving equation. If students get stuck at any one out of the three stages, the word problems would not be solved proficiently.

From the above mentioned discussion on the learning matters, it could be seen that there is much attention paid to the students' specific challenges in certain topics of algebra learning. The challenges' explosion could be incurred by all the categories already discussed such as algebra content, cognitive gap, and teaching issues. Thus, the discussed learning matters are consequential phenomena caused by the structural nature of algebra, cognitive gap, and the absence of teaching concerns.

## TRANSITION KNOWLEDGE

There are two kinds of knowledge emerging from the literature to address the knowledge required in the transition from arithmetic to algebra: adaptation to a milieu and social interactions; and the adjustments.

### Adaptation to a Milieu and Social Interaction

Sadovsky and Sessa (2005) aimed to give an account of the emergence of knowledge pertaining to the transition from arithmetic to algebra in the course of an algebra learning classroom with two kinds of interaction such as the adidactic interaction between each student and a given problem, and the adidactic interaction of each student with the procedures of others. It was assumed by them that the processes of adaptation to a milieu and the social interaction were crucial for the transition from arithmetic to algebra.

## Adjustments

Kilpartick, Swafford, & Findell (2001) revealed that students needed to make many adjustments in the transition from arithmetic to algebra even for excellent students in arithmetic. The assumption was supported by many examples provided by their analysis. For example, an adjustment from answers orientation to relations orientation was illustrated from the following statement "elementary school arithmetic tends to be heavily answer- oriented and does not focus on the representation of relations" (p. 261). A specific example provided by them was that students always assumed 8+5 as a computing signal and typically wrote 13 for the number sentence 8+5=_+9 by evaluating it instead of the correct answer 4. Another adjustment from undoing operation to expressing equation was exemplified by a problem solving "when 3 is added to 5 times a certain number, the sum is 38; find the number" (p. 262). The arithmetic method is undoing in reverse order (subtract 3 from 38 and then divide by 5), while algebraic way is to represent the relationships by the stated operation: 5x+3=38. Therefore, the different methods require the students to make certain adjustments.

Based on the idea of adjustments, Kieran (2004: pp. 140-141) defined five kinds of adjustments: 1) a focus on relations and not merely on the calculation of a numerical answer; 2) a focus on operations as well as their inverses, and on the related idea of doing/undoing; 3) a focus on both representing and solving a problem rather than on merely solving it; 4) a focus on both numbers and letters, rather than on numbers alone including: working with letters that may at times be unknowns, variables, or parameters; accepting unclosed literal expressions as responses; comparing expressions for equivalence based on properties rather than on numerical evaluation; and 5) a refocusing of the meaning of the equal sign. The five adjustments are not only the knowledge we should know about students' transition from arithmetic to algebra but also the guidance for teaching to prevent or treat students' difficulties in algebra learning.

In fact, the transition knowledge generated from the literature is taken as the systemization of main challenges students might encounter and the suggestions of overcoming their difficulties in the algebra learning. Therefore, the knowledge is also regarded as a remedy for the scarcity of teaching orientation.

## CONCLUSIONS

In this literature review, it is attempted to re-conceptualize much of the existing algebra researches by focusing on the challenges that students might confront with in learning algebra from the perspectives of algebra content, cognitive gap, teaching issues, learning matters, and transition knowledge. The perspectives adopted in this paper could serve as a framework of better understanding students' difficulties in their algebra learning.

In addition, from the critical perspective of reviewing the existing researches, it could be found that the existing researches are conducted from the static angle to examine the causes resulting in students' difficulties in algebra learning. In another word, there is a lack of process analysis of students' going through the transition from arithmetic to algebra. Without the undoing of the process, we could not perceive the circumstances that the students could struggle with so well that we could not provide apt strategies to prevent or remedy the difficulties (Wang, 2014: p. 2) in a systemic way even though we know the existing difficulties and their causes.

# REFERENCES

1. Banerjee, R., & Subramaniam, K. (2012). Evolution of a Teaching Approach for BeginningAlgebra. Educational Studies in Mathematics, 80, 351-367.http://dx.doi.org/10.1007/s10649-011-9353-y

2. Behr, M., Erlwanger, S., & Nichols, E. (1980). How Children View the Equals Sign. Mathematics Teaching, 92, 13-18.

3. Booth, L. R. (1984). Algebra: Children's Strategies and Errors. Windsor, UK: NFER-Nelson.

4. Booth, L. R. (1988). Children's Difficulties in Beginning Algebra. In A. F. Coxford (Ed.), The Ideas of Algebra, K-12 (1988 Yearbook, pp. 20-32). Reston, VA: National Council of Teachers of Mathematics.

5. Cai, J., & Moyer, J. (2008). Developing Algebraic Thinking in Earlier Grades: Someinsights from International Comparative Studies. In C. Greenes, & R. Rubenstein (Eds.), Algebra and Algebraic Thinking in School Mathematics (70th Yearbook of the National Council of Teachers of Mathematics, pp.169-180). Reston, VA: NCTM.

6. Cai, J., Lew, H. C, Morris, A., Mover, J. C, Ng, S. F., & Schmittau, J. (2004). The Development of Students' Algebraic Thinking in Earlier Grades: A Cross-Cultural Comparative Perspective. Paper Presented at the Annual Meeting of the American Educational Research Association, San Diego, CA.

7. Carry, L. R., Lewis, C., & Bernard, J. (1980). Psychology of Education Solving: An Information Processing Study. Austin: University of Texas at Austin, Department of Curriculum and Instruction.

8. Chaiklin, S. (1989). Cognitive Studies of Algebra Problem Solving and Learning. In S. Wagner, & Kieran (Eds.), Research Issue in Learning and Teaching of Algebra (pp. 93-114). Reston, VA: National Council of Teachers of Mathemaics; Hillsdale, NJ: Lawrence Erlbaum.

9. Davis, R. B. (1975). Cognitive Processes Involved in Solving Simple Algebraic Equations. Journal of Children's Mathematical Behaviour, 1, 7-35.

10. Falkner, K., Levi, L., & Carpenter, T. P. (1999). Children's Understanding of Equality Foundation for Algebra. Teaching Children Mathematics, 6, 232-237.

11. Filloy, E., & Rojano, T. (1989). Solving Equations: The Transition from Arithmetic to Algebra. For the Learning of Mathematics, 9, 19-25.

12. Fischbein, E., & Barash, A. (1993). Algorithmic Models and Their Misuse in Solving Algebraic Problems. Proceedings of PME 17, 1, 162-172.

13. Freiman, V., & Lee, L. (2004). Tracking Primary Students' Understanding of Equal Sign. In M. Hoines, & A. Fuglestad (Eds.), Proceedings of the 28th Conference of the International Group for the Psychology of Mathematics Education (pp. 415-422). Bergen: PME.

14. Greeno, J. G. (1982). A Cognitive Learning Analysis of Algebra. The Annual Meeting of the American Educational Research Association, Boston, MA.

15. Herscovics, N. (1989). Cognitive Obstacles Encountered in the Learning of Algebra. In S. Wagner, & C. Kieran (Eds.), Research Issues in the Learning and Teaching of Algebra (pp. 60-86). Reston, VA: National Council of Teachers of Mathematics; Hillsdale, NJ: Lawrence Erlbaum.

16. Herscovics, N., & Linchevski, L. (1994). A Cognitive Gap between Arithmetic and Algebra. Educational Studies in Mathematics, 27, 59-78. http://dx.doi.org/10.1007/BF01284528

17. Hoz, R., & Harel, G. (1989). The Facilitating Role of Table Form in Solving Algebra Speed Problems: Real or Imaginary? In G. Vergnaud, J. Rogalski, & M. Artigue (Eds.), Proceeding of the 13th International Conference for the Psychology of Mathematics Education (pp. 123-130). Paris: G. R. Didactique, CNRS.

18. Khng, K. H., & Lee, K. (2009). Inhibiting Interference from Prior Knowledge: Arithmetic Intrusions in Algebra Word Problem Solving. Learning and Individual Differences, 19, 262-268. http://dx.doi.org/10.1016/j.lindif.2009.01.004

19. Kieran, C. (1992). The Learning and Teaching of School Algebra. In D. Grouws (Ed.), Handbook of Research on Mathematics Teaching and Learning (pp. 390-419). New York: Macmillan Publishing Company.

20. Kieran, C. (2004). Algebraic Thinking in the Early Grades: What Is It? The Mathematics Educator, 8, 139-151.

21. Kiichemann, D. (1981). Algebra. In K. M. Hart (Ed.), Children's Understanding of Mathematics (pp. 11-16). London: John Murray.

22. Kilpatrick, J., Swafford, J., & Findell, B. (2001). Adding It Up: Helping Children Learn Mathematics. Washington, DC: National Academy Press.

23. Li, X., Ding, M., Capraro, M. M., & Capraro, R. M. (2008). Sources of Differences in Children's Understandings of Mathematical Equality: Comparative Analysis of Teacher Guides and Student Texts in China and the United States. Cognition and Instruction, 26, 195-217. http://dx.doi.org/10.1080/07370000801980845

24. Linchevski, L., & Herscovics, N. (1996). Crossing the Cognitive Gap between Arithmetic and Algebra: Operating on the Unknown in the Context of Equations. Educational Studies in Mathematics, 30, 39-65. http://dx.doi.org/10.1007/BF00163752

25. Napaphun, V. (2012). Relational Thinking: Learning Arithmetic in Order to Promote Algebraic Thinking. Journal of Science and Mathematics Education in Southeast Asia, 35, 84-101.

26. Ng, S. F., & Lee, K. (2009). The Model Method: Singapore Children's Tool for Representing and Solving Algebraic Word Problems. Journal for Research in Mathematics Education, 40, 282-313.

27. Rachlin, S. L. (1989). Using Research to Design a Problem-Solving Approach for Teaching Algebra. In S. T. Ong (Ed.), Proceedings of the 4th Southeast Asian Conference on Mathematical Education (pp. 156-161). Singapore: Singapore Institute of Education.

28. Radford, L. (2012). Early Algebraic Thinking Epistemological, Semiotic, and Developmental Issues. 12th International Congress on Mathematical Education, Seoul, South Korea.

29. Radford, L., & Puig, L. (2007). Syntax and Meaning as Sensuous, Visual, Historical forms of Algebraic Thinking. Educational Studies in Mathematics, 66, 145-164. http://dx.doi.org/10.1007/s10649-006-9024-6

30. Reed, S. K. (1987). A Structure-Mapping Model for Word Problems. Journal of Experimental Psychology: Learning, Memory and Cognition, 13, 124-139. http://dx.doi.org/10.1037/0278-7393.13.1.124

31. Reed, S. K., Dempster, A., & Ettinger, M. (1985). The Usefulness of Analogous Solution for Solving Algebra Word Problems. Journal of Experimental Psychology: Learning, Memory and Cognition, 11, 106-125. http://dx.doi.org/10.1037/0278-7393.11.1.106

32. Sadovsky, P., & Sessa, C. (2005). The Adidactic Interaction with the Procedures of Peers in the Transition from Arithmetic to Algebra: A Milieu for the Emergence of New Questions. Educational Studies in Mathematics, 59, 85-112. http://dx.doi.org/10.1007/s10649-005-5886-2

33. Sfard, A. (1991). On the Dual Nature of Mathematical Conceptions: Reflections on Processes and Objects as Different Sides of the Same Coin. Educational Studies in Mathematics, 22, 1-36. http://dx.doi.org/10.1007/BF00302715

34. Sfard, A., & Linchevski, L. (1993). Processes without Objects—The Case of Equations and Inequalities. The Special Issue of del Seminario Matematico de U'Universita edel Politecnico di Torino.

35. Sfard, A., & Linchevski, L. (1994). The Gains and the Pitfalls of Reification—The Case of Algebra. Educational Studies in Mathematics, 26, 191-228.http://dx.doi.org/10.1007/BF01273663

36. Wang, X. (2014). The Transition from Arithmetic to Algebra: Cognitive Gap, Pre-algebraic Conceptualization, and Teacher Preparation. Edmonton: University of Alberta. (Unpublished Essay).

37. Welder, R. M. (2012). Improving Algebra Preparation: Implications from Research on Student Misconceptions and Difficulties. School Science and Mathematics, 112, 255-264.http://dx.doi.org/10.1111/j.1949-8594.2012.00136.x

38. Wenger, R. (1987). Cognitive Science and Algebra Learning. In A. Schoenfeld (Ed.), Cognitive Science and Mathematics Education (pp. 217-251). Hillsdale, NJ: Lawrence Erlbaum.

## CITATION

Wang, X. (2015) The Literature Review of Algebra Learning: Focusing on the Contributions to Students' Difficulties. Creative Education, 6, 144-153. doi: 10.4236/ce.2015.62013.

# Geometric Analogy and Products of Vectors in n Dimensions

## *Leonardo Simal Moreira*

UniFOA—Centro Universitário de Volta Redonda, Volta Redonda, Brazil

## ABSTRACT

The cross product in Euclidean space $IR^3$ is an operation in which two vectors are associated to generate a third vector, also in space $IR^3$. This product can be studied rewriting its basic equations in a matrix structure, more specifically in terms of determinants. Such a structure allows extending, for analogy, the ideas of the cross product for a type of the product of vectors in higher dimensions, through the systematic increase of the number of rows and columns in determinants that constitute the equations. So, in a n-dimensional space with Euclidean norm, we can associate n − 1 vectors and to obtain an n-th vector, with the same geometric characteristics of the product in three dimensions. This kind of operation is also a geometric interpretation of the product defined by Eckman [1]. The same analogies are also useful in the verification of algebraic properties of such products, based on known properties of determinants.

## INTRODUCTION

In the Euclidean space $IR^3$, the cross product of two vectors u and v is the vector designated by the symbol [uv], and defined for the following conditions [2]:

a) The norm of vector [uv], symbolized for $\|[uv]\|$, is given for

$$\|[uv]\| = |u||v|k ,$$

(1)

Where $k = sen\alpha$, being $\alpha$ the angle between the vectors u and v.

b) The vector [uv] is perpendicular simultaneously to the vectors u and v:

$$[uv] \cdot u = 0 ,$$

(2)

$$[uv] \cdot v = 0 .$$

(3)

As a consequence of b), [uv] is the normal vector to the plane defined for the vectors u and v (Figure 1), if these are linearly independent vectors. Considering $[uv] = (p,q,r)$, then $\beta: px + qy + rz + c = 0$, where $c = -pa_1 - qa_2 - ra_3$, represents the equation of the plane $\beta$ in a Cartesian coordinate system $A(a_1, a_2, a_3)$ is a point in $IR^3$ and $A \in \beta$).

If u and v are linearly dependent vectors, then

$$[uv] = 0 ,$$

(4)

where the symbol 0 represents the null vector.

c) The vector $[uv]$ is oriented in relation to the vectors u and v just as, in right-handed coordinate system, the z-axis it is oriented in relation to the x-axis and y-axis.

d) The volume $V_3$ of parallelepiped defined for the vectors u, v and $[uv]$ is the square of the number $\|[uv]\|$ (Figure 2):

$$V_3 = \left\|\left[uv\right]\right\|^2$$

(5)

The equalities (2), (3) and (5) are equivalent to those given in a Definition 1 found in [3].

In this paper, it is shown that it is possible, through simple analogies with the case in the space $IR^3$, to extend the ideas of the cross product to the space $IR^4$, and more generally, to the space $IR^n$. The characteristics of the cross product in $IR^3$ are maintained in higher dimensions.

## MATRIX STRUCTURE OF $[uv]$

The initial reasoning for the extension of the ideas of the cross product is the fact that their basic expressions can be represented in the form of determinants.

In an orthogonal coordinate system, representing the vectors u and v in terms of 3-tuples $u = (u_1, u_2, u_3)$ and $v = (v_1, v_2, v_3)$, the vector $[uv]$ can be obtained starting from the development of the symbolic determinant

$$[uv] = \begin{vmatrix} \hat{e}_1 & \hat{e}_2 & \hat{e}_3 \\ u_1 & u_2 & u_3 \\ v_1 & v_2 & v_3 \end{vmatrix},$$

(6)

where $\hat{e}_1 = (1,0,0), \hat{e}_2 = (0,0,1), \hat{e}_3 = (0,0,1)$ are the vectors of orthonormal basis in $IR^3$.

The development of the Equation (6) leads to the vector form:

$$[uv] = \begin{vmatrix} u_2 & u_3 \\ v_2 & v_3 \end{vmatrix} \hat{e}_1 - \begin{vmatrix} u_1 & u_3 \\ v_1 & v_3 \end{vmatrix} \hat{e}_2 + \begin{vmatrix} u_1 & u_2 \\ v_1 & v_2 \end{vmatrix} \hat{e}_3,$$

(7)

and the norm of vector [uv] is calculated with the definition of Euclidean norm, resulting in

$$\|[uv]\| = \left( \begin{vmatrix} u_2 & u_3 \\ v_2 & v_3 \end{vmatrix}^2 + \begin{vmatrix} u_1 & u_3 \\ v_1 & v_3 \end{vmatrix}^2 + \begin{vmatrix} u_1 & u_2 \\ v_1 & v_2 \end{vmatrix}^2 \right)^{\frac{1}{2}},$$

(8)

an equivalent format to

$$\|[\boldsymbol{uv}]\| = \begin{vmatrix} \begin{vmatrix} u_2 & u_3 \\ v_2 & v_3 \end{vmatrix} & (-1)\begin{vmatrix} u_1 & u_3 \\ v_1 & v_3 \end{vmatrix} & \begin{vmatrix} u_1 & u_2 \\ v_1 & v_2 \end{vmatrix}^{\frac{1}{2}} \\ u_1 & u_2 & u_3 \\ v_1 & v_2 & v_3 \end{vmatrix}.$$

(9)

In Equation (1),

$$k = \operatorname{sen}\alpha = \begin{vmatrix} 1 & \cos\alpha \\ \cos\alpha & 1 \end{vmatrix}^{\frac{1}{2}}, (0 \leq \alpha \leq \pi),$$

(10)

and combining the Equations (9) and (10), we obtain

$$\begin{vmatrix} \begin{vmatrix} u_2 & u_3 \\ v_2 & v_3 \end{vmatrix} & (-1)\begin{vmatrix} u_1 & u_3 \\ v_1 & v_3 \end{vmatrix} & \begin{vmatrix} u_1 & u_2 \\ v_1 & v_2 \end{vmatrix}^{\frac{1}{2}} \\ u_1 & u_2 & u_3 \\ v_1 & v_2 & v_3 \end{vmatrix}$$

$$= \|\boldsymbol{u}\|\|\boldsymbol{v}\| \begin{vmatrix} 1 & \cos\alpha \\ \cos\alpha & 1 \end{vmatrix}^{\frac{1}{2}}$$

(11)

Equation (11) will be used as starting point for the analogies developed in the remaining of this work.

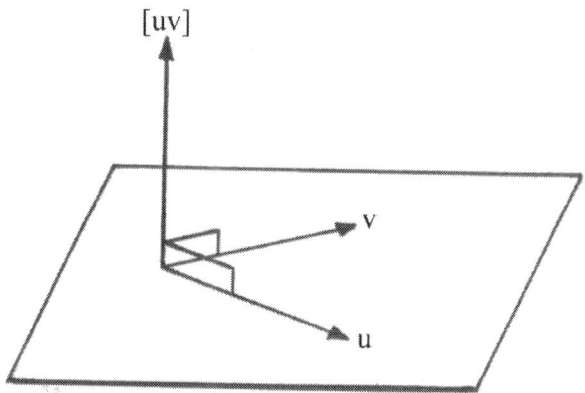

**Figure 1**: [uv] is the normal vector to the plane defined for the vectors u and v.

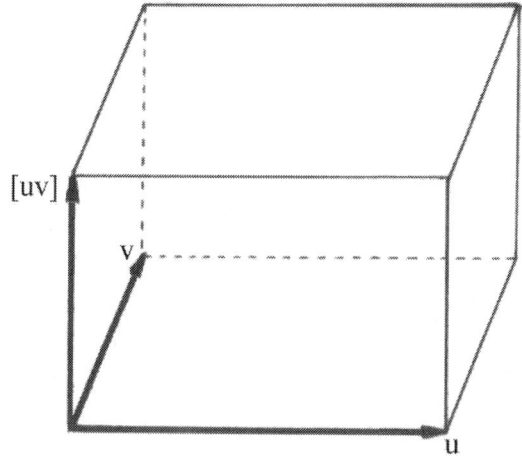

**Figure 2**: Parallelepiped defined for the vectors u, v and [uv].

## EXTENSION OF THE CROSS PRODUCT TO THE EUCLIDEAN SPACE $IR^4$

Consider three vectors in Euclidean space $IR^4$, represented in terms of quadruples $v_1 = (v_{11}, v_{12}, v_{13}, v_{14})$, $v_2 = (v_{21}, v_{22}, v_{23}, v_{24})$ and $v_3 = (v_{31}, v_{32}, v_{33}, v_{34})$. Let $\hat{e}_1 = (1,0,0,0), \hat{e}_2 = (0,1,0,0), \hat{e}_3 = (0,0,1,0)$ and $\hat{e}_4 = (0,0,0,1)$ be the vectors of orthonormal basis in $IR^4$.

It is possible to develop an equivalent product to (1), through simple extension of ideas and increase of dimensions. In space $IR^3$, two vectors u and v generate a third vector whose norm is proportional to the product of the norms of the generating vectors, being the proportionality constant related to the angle between u and v. In space $IR^4$, three vectors $v_1, v_2$ and $v_3$ generate a fourth vector whose norm is proportional to the product of the norms of the generating vectors, being the proportionality constant related to the angles between the vectors $v_1$ and $v_2, v_1$ and $v_3, v_2$ and $v_3$.

In symbolic terms, this product of vectors in Euclidean space $IR^4$ is obtained from the development of the determinant

$$h = \left[ v_1 v_2 v_3 \right] = \begin{vmatrix} \hat{e}_1 & \hat{e}_2 & \hat{e}_3 & \hat{e}_4 \\ v_{11} & v_{12} & v_{13} & v_{14} \\ v_{21} & v_{22} & v_{23} & v_{24} \\ v_{31} & v_{32} & v_{33} & v_{34} \end{vmatrix},$$

(12)

so that

$$|h| = |v_1||v_2||v_3|\,\overline{k}\;,$$

(13)

with

$$\overline{k} = \begin{vmatrix} 1 & \cos\alpha_{12} & \cos\alpha_{13} \\ \cos\alpha_{21} & 1 & \cos\alpha_{23} \\ \cos\alpha_{31} & \cos\alpha_{32} & 1 \end{vmatrix}^{\frac{1}{2}},$$

(14)

$$\begin{vmatrix} \begin{vmatrix} v_{12} & v_{13} & v_{14} \\ v_{22} & v_{23} & v_{24} \\ v_{32} & v_{33} & v_{34} \end{vmatrix} & (-1)\begin{vmatrix} v_{11} & v_{13} & v_{14} \\ v_{21} & v_{23} & v_{24} \\ v_{31} & v_{33} & v_{34} \end{vmatrix} & \begin{vmatrix} v_{11} & v_{12} & v_{14} \\ v_{21} & v_{22} & v_{24} \\ v_{31} & v_{32} & v_{34} \end{vmatrix} & (-1)\begin{vmatrix} v_{11} & v_{12} & v_{13} \\ v_{21} & v_{22} & v_{23} \\ v_{31} & v_{32} & v_{33} \end{vmatrix} \\ v_{11} & v_{12} & v_{13} & v_{14} \\ v_{21} & v_{22} & v_{23} & v_{24} \\ v_{31} & v_{32} & v_{33} & v_{34} \end{vmatrix}^{\frac{1}{2}} = |v_1||v_2||v_3| \begin{vmatrix} 1 & \cos\alpha_{12} & \cos\alpha_{13} \\ \cos\alpha_{21} & 1 & \cos\alpha_{23} \\ \cos\alpha_{31} & \cos\alpha_{32} & 1 \end{vmatrix}^{\frac{1}{2}}$$

(15)

and the conditions $\begin{pmatrix} \alpha_{12} + \alpha_{13} + \alpha_{21} \le 2\pi, \\ \alpha_{12} \le \alpha_{13} + \alpha_{21}, \\ \alpha_{13} \le \alpha_{12} + \alpha_{21}, \\ \alpha_{21} \le \alpha_{12} + \alpha_{13}, \end{pmatrix}$.

The equal sign in the conditions on the angles, given in (14), is justified for the case of coplanar vectors.

In Equation (14), $\cos\alpha_{ij} = \dfrac{v_i.v_j}{|v_i||v_j|}$ represents the angle between two of the generating vectors of h, and naturally $\cos a_{ij} = \cos a_{ij}$, so that $\overline{k}^2$ is the determinant of a symmetric matrix.

The equivalent in space $IR^4$ of Equation (11) is (see the Equation (15) below):

The characteristics of the product $[uv]$ in space $IR^3$ are conserved for h in space $IR^4$:

a) The norm of h is proportional to the product $|v_1 \| v_2 \| v_3|$. It is sufficient to develop the determinants in Equation (15) to verify the identity.

b) The vector h is perpendicular to each one of the vectors $v_1, v_2$ and $v_3$. The term "perpendicular" should be interpreted here as only in the sense that the scalar product $h.v_i$ results null.

## Proof

The elements of the 1st row of the determinant that represents the norm of h are the same values as their own cofactors. It is known that the sum of the products of the elements of a row for the cofactors of the elements corresponding of other row (inner product) in a determinant results in zero (Cauchy's Determinant Theorem), that is, $h.v_1 = h.v_2 = h.v_3 = 0$.

It is also noted that h is the normal vector to the hyperplane that contains $v_1, v_2$ and $v_3$. Being $h = h_1 \hat{e}_1 + h_2 \hat{e}_2 + h_3 \hat{e}_3 + h_4 \hat{e}_4$, then $H : h_1 x_1 + h_2 x_2 + h_3 x_3 + h_4 x_4 + \bar{c} = 0$, where $\bar{c} = -h_1 a_1 - h_2 a_2 - h_3 a_3 - h_4 a_4$, represents the Cartesian equation of hyperplane H $A(a_1, a_2, a_3, a_4)$ is a point in $IR^4$ and $A \in H$ ).

c) The vector h is oriented in relation to the vectors $v_1, v_2$ and $v_3$ just as the vector $-\hat{e}_4$ in relation to $\hat{e}_1, \hat{e}_2$ and $\hat{e}_3$.

d) The content of parallelotope defined for the vectors $v_1, v_2, v_3$ and h is the square of number $|h|$.

## *Proof*

With effect, the determinant to the left in Equation (15) represents the number $|h|$. In this way, $|h|^2$ is the determinant whose rows are formed by the vectors $h, v_1, v_2$ and $v_3$, representing the content of parallelotope (4-parallelepiped) that has the four vectors as edges linearly indepen-dents [4].

## PRODUCT OF N – 1 VECTORS IN EUCLIDEAN SPACE IR$^N$

Consider n – 1 vectors in Euclidean space $IR^n$, represented in terms of n-tuples, such that

$$v_1 = \left(v_{11}, v_{12}, \cdots, v_{1,n}\right), v_2 = \left(v_{21}, v_{22}, \cdots, v_{2,n}\right)$$
$$, \cdots, v_{n-1} = \left(v_{n-1,1}, v_{n-1,2}, \cdots, v_{n-1,n}\right)$$

The product $H = [v_1 v_2 \cdots v_{n-1}]$ in space $IR^n$ is a vector perpendicular simultaneously to all the $v_i (1 \leq i \leq n-1)$ and whose norm is given by the formula

$$|H| = |v_1||v_2|\cdots|v_{n-1}|K ,$$

$$(16)$$

with

$$
K = \begin{vmatrix}
1 & \cos\alpha_{1,2} & \cdots & \cos\alpha_{1,n-1} \\
\cos\alpha_{2,1} & 1 & \cdots & \cos\alpha_{2,n-1} \\
\vdots & \vdots & \ddots & \vdots \\
\cos\alpha_{n-1,1} & \cos\alpha_{n-1,2} & \cdots & 1
\end{vmatrix}^{\frac{1}{2}}.
$$

(17)

It is observed that this form is equivalent to the products of vectors defined by [1], and cited in [5,6], namely (using the same symbols as in [6]), that a cross product satisfies the axioms:

$$
\langle P(a_1,\cdots,a_r),a_i \rangle = 0, (1 \le i \le r),
$$

(A1)

$$
\| P(a_1,\cdots,a_r) \|^2 = \det(\langle a_i,a_j \rangle),
$$

(A2)

Where $\|a\|^2 = \langle a,a \rangle$.

These preliminary definitions can be formalized starting from the following proposition.

## Proposition

Let $n-1$ vectors be in space $IR^n$, with inner product and Euclidean norm. Consider also that the vectors are represented by n-tuples such that

$$v_1 = \left( v_{11}, v_{12}, \cdots, v_{1,n} \right), v_2 = \left( v_{21}, v_{22}, \cdots, v_{2,n} \right)$$
$$\cdots, v_{n-1} = \left( v_{n-1,1}, v_{n-1,2}, \cdots, v_{n-1,n} \right)$$

Being $\cos a_{ij}$ the angle between the i-th vector $v^i$ and the j-th vector $v_j$, the following equality is true (see the Equation (18) below):

$$\left\| \begin{matrix} v_{12} & v_{13} & \cdots & v_{1n} \\ v_{22} & v_{23} & \cdots & v_{2n} \\ \vdots & \vdots & \ddots & \vdots \\ v_{n-1,2} & v_{n-1,3} & \cdots & v_{n-1,n} \\ & v_{11} & \vdots & \\ & v_{n-1,1} & & \end{matrix} \right| (-1) \left| \begin{matrix} v_{11} & v_{13} & \cdots & v_{1n} \\ v_{21} & v_{23} & \cdots & v_{2n} \\ \vdots & \vdots & \ddots & \vdots \\ v_{n-1,1} & v_{n-1,3} & \cdots & v_{n-1,n} \\ & v_{12} & \vdots & \\ & v_{n-1,2} & & \end{matrix} \right| \cdots (-1)^{1+j} \left| \begin{matrix} v_{11} & v_{12} & \cdots & v_{1,n-1} \\ v_{21} & v_{22} & \cdots & v_{2,n-1} \\ \vdots & \vdots & \ddots & \vdots \\ v_{n-1,1} & v_{n-1,2} & \cdots & v_{n-1,n-1} \\ & v_{1n} & \vdots & \\ & v_{n-1,n} & & \end{matrix} \right|^{\frac{1}{2}}$$

$$= |v_1||v_2| \cdots |v_{n-1}| \left| \begin{matrix} 1 & \cos \alpha_{12} & \cdots & \cos \alpha_{1,n-1} \\ \cos \alpha_{21} & 1 & \cdots & \cos \alpha_{2,n-1} \\ \vdots & \vdots & \ddots & \vdots \\ \cos \alpha_{n-1,1} & \cos \alpha_{n-1,2} & \cdots & \cos \alpha_{n-1,n-1} \end{matrix} \right|^{\frac{1}{2}} \tag{18}$$

## Proof

Consider $n - 1$ unit vectors $u_i (1 \le i \le n-1)$ in space $\mathbb{R}^n$, with inner product and Euclidean norm. Consider also that each $u_i$ represents the unit vector in the same direction of $v_{iv}$ given in the Equation (18), so that

$$u_i = \frac{v_i}{|v_i|}. \tag{19}$$

If the unit vectors are represented by n-tuples such that

$u_i = (u_{11}, u_{12}, \cdots u_{1n}), u_2 = (u_{21}, u_{22}, \cdots, u_{2n})$ being $u_i.u_j$ the inner product
$, \cdots, u_{n-1} = (u_{n-1,1}, u_{n-1,2}, \cdots, u_{n-1,n})$

between the i-th unit vector $u_i$ and the j-th unit vector $u_j$, can be grouped, based on the properties presented in (A2), the components of $u_i$ in the following identity, which is true for values of $n \geq 3$:

$$\begin{vmatrix} u_{12} & u_{13} & \cdots & u_{1n} \\ u_{22} & u_{23} & & u_{2n} \\ \vdots & & \ddots & \vdots \\ u_{n-1,2} & u_{n-1,3} & & u_{n-1,n} \\ & u_{11} & & \\ & \vdots & & \\ & u_{n-1,1} & & \end{vmatrix} (-1) \begin{vmatrix} u_{11} & u_{13} & \cdots & u_{1n} \\ u_{21} & u_{23} & & u_{2n} \\ \vdots & \vdots & & \vdots \\ u_{n-1,1} & u_{n-1,3} & \cdots & u_{n-1,n} \\ & u_{12} & & \\ & \vdots & & \\ & u_{n-1,2} & & \end{vmatrix} \cdots (-1)^{1+j} \begin{vmatrix} u_{11} & u_{12} & \cdots & u_{1,n-1} \\ u_{21} & u_{22} & & u_{2,n-1} \\ \vdots & & \ddots & \vdots \\ u_{n-1,1} & u_{n-1,2} & & u_{n-1,n-1} \\ & u_{1n} & & \\ & \vdots & & \\ & u_{n-1,n} & & \end{vmatrix}$$

$$= \begin{vmatrix} 1 & u_1 \cdot u_2 & \cdots & u_1 \cdot u_{n-1} \\ u_2 \cdot u_1 & 1 & \cdots & u_2 \cdot u_{n-1} \\ \vdots & & \ddots & \vdots \\ u_{n-1} \cdot u_1 & u_{n-1} \cdot u_2 & \cdots & 1 \end{vmatrix} \qquad (20)$$

Starting from Equation (20), Equation (18) can be demonstrated. With effect, multiplying both members of (20) for $(|v_1| |v_2| \cdots |v_{n-1}|)^2$, the determinant to the left will have their rows orderly and appropriately multiplied by each one of $|v_i|$, and since $|v_i| u_i = v_i$, is obtained the corresponding determinant of Equation (18).

Representing, for convenience,

$$\begin{vmatrix} v_{12} & v_{13} & \cdots & v_{1n} \\ v_{22} & v_{23} & \cdots & v_{2n} \\ \vdots & \vdots & \ddots & \vdots \\ v_{n-1,2} & v_{n-1,3} & \cdots & v_{n-1,n} \\ & v_{11} & & \\ & \vdots & & \\ & v_{n-1,1} & & \end{vmatrix} (-1) \begin{vmatrix} v_{11} & v_{13} & \cdots & v_{1n} \\ v_{21} & v_{23} & \cdots & v_{2n} \\ \vdots & \vdots & \ddots & \vdots \\ v_{n-1,1} & v_{n-1,3} & \cdots & v_{n-1,n} \\ & v_{12} & & \\ & \vdots & & \\ & v_{n-1,2} & & \end{vmatrix} \cdots (-1)^{1+j} \begin{vmatrix} v_{11} & v_{12} & \cdots & v_{1,n-1} \\ v_{21} & v_{22} & \cdots & v_{2,n-1} \\ \vdots & \vdots & \ddots & \vdots \\ v_{n-1,1} & v_{n-1,2} & \cdots & v_{n-1,n-1} \\ & v_{1n} & & \\ & \vdots & & \\ & v_{n-1,n} & & \end{vmatrix} = \begin{Vmatrix} v_{ik} \cdot k \neq j \\ v_{ij} \end{Vmatrix}$$

$(1 \leq i \leq n-1, 1 \leq j \leq n. 1 \leq k \leq n)$ \hspace{4cm} (21) we have that:

$$\left(|v_1||v_2|\cdots|v_{n-1}|\right)^2\begin{Vmatrix}u_{ik}, k \neq j \\ u_{ij}\end{Vmatrix} = \begin{Vmatrix}v_{ik}, k \neq j \\ v_{ij}\end{Vmatrix}.$$

(22)

In relation to the determinant to the right in Equation (20), it is sufficient to observe that $|u_i| = 1$, therefore $\cos\alpha_{ij} = u_i.u_j$, that is:

$$\begin{vmatrix} 1 & u_1 \cdot u_2 & \cdots & u_1 \cdot u_{n-1} \\ u_2 \cdot u_1 & 1 & \cdots & u_2 \cdot u_{n-1} \\ \vdots & \vdots & \ddots & \vdots \\ u_{n-1} \cdot u_1 & u_{n-1} \cdot u_2 & \cdots & 1 \end{vmatrix}$$

$$= \begin{vmatrix} 1 & \cos\alpha_{12} & \cdots & \cos\alpha_{1,n-1} \\ \cos\alpha_{21} & 1 & \cdots & \cos\alpha_{2,n-1} \\ \vdots & \vdots & \ddots & \vdots \\ \cos\alpha_{n-1,1} & \cos\alpha_{n-1,2} & \cdots & \cos\alpha_{n-1,n-1} \end{vmatrix}$$

(23)

With such considerations, it is demonstrated that

$$\begin{Vmatrix}v_{ik}, k \neq j \\ v_{ij}\end{Vmatrix} = \left(|v_1||v_2|\cdots|v_{n-1}|\right)^2$$

$$\times \begin{vmatrix} 1 & \cos\alpha_{12} & \cdots & \cos\alpha_{1,n-1} \\ \cos\alpha_{21} & 1 & \cdots & \cos\alpha_{2,n-1} \\ \vdots & \vdots & \ddots & \vdots \\ \cos\alpha_{n-1,1} & \cos\alpha_{n-1,2} & \cdots & \cos\alpha_{n-1,n-1} \end{vmatrix},$$

(24)

and the square root of Equation (23) shows that Equation (18) is true.

Equation (18) is the equivalent n-dimensional of the Equations (11) and (15), validating the extension of cross product. The geometric properties of H are conserved in n dimensions:

a) The norm of H is proportional to the product $|v_1||v_2|\cdots|v_{n-1}|$, being the proportionality constant K associated to the angles between the vectors $v_1$.

*Proof*

The proof consists of the own demonstration of the Equation (18).

b) The vector H is "perpendicular" to each one of the vectors $v_1, v_2, \cdots, v_{n-1}$.

*Proof*

The elements of the 1st row of the determinant that represents the norm of H are the same values as their own cofactors. In agreement with Cauchy's Determinant Theorem, the sum of the products of the elements of a row for the cofactors of the elements corresponding of another row (inner product) in a determinant results in zero, that is,

$$H.v_1 = H.v_2 = \cdots = H.v_{n-1} = 0.$$

It is also noted that H is the normal vector to the hyperplane that contains $v_1, v_2, \cdots, v_{n-1}$. Being

$$H = H_1\hat{e}_1 + H_2\hat{e}_2 + \cdots + H_n\hat{e}_n \text{ , then}$$

$H_n : H_1 x_1 + H_2 x_2 + \cdots + H_n x_n + C = 0$, where

$C = -H_1 a_1 - H_2 a_2 - \cdots - H_n a_n$, represents the Cartesian equation of hyperplane $H_n$ ( $A(a_1, a_2 \cdots, a_n)$ is a point in $IR^n$ and $A \in H^n$ ).

c) The vector H is oriented in relation to the vectors $v_1, v_2, \cdots, v_{n-1}$ just as the vector $(-1)^{n+1} \hat{e}_n$ is oriented in relation to $\hat{e}_1, \hat{e}_2, \cdots, \hat{e}_{n-1}$.

d) The content of parallelotope defined for the vectors $v_1, v_2, \cdots, v_{n-1}$ and H is the square of number $|H|$.

*Proof*

The determinant to the left in Equation (18) represents the number $|H|$. In this way, $|H|^2$ is the determinant whose rows are formed by the vectors $H, v_1, v_2, \cdots, v_{n-1}$, representing the content of parallelotope (n-parallelepiped) that has the n vectors as edges linearly independents [4].

## CONCLUSIONS

The possibility to represent the equations of the definition of cross product in the space $IR^3$ in terms of determinants allows the extension of the concept of the product of vectors for higher dimensions, systematically increasing rows and columns to the determinants.

Through basic properties of determinants, it is shown that the characteristics of the cross product are conserved in n dimensions, for any value of n, since such properties are not modified by the increment or decrease of rows and columns to these determinants.

Other geometric properties can be verified, as the relationship be-
tween the cross product and area, because just as the number $|[uv]|$
is related to areas of triangles and parallelograms, the number $|H|$ is
related to contents of simplex and parallelotopes, in an equivalent way
to Cayley-Menger determinant [7, 8].

Although this work has given emphasis to the geometric properties of
the product of vectors in the space $IR^n$, it indirectly shows that their
algebraic properties are also similar to those valid ones in space $IR^3$,
for instance:

(C1) If $w_i$ is any vector in space $IR^n$ for

$i = 1, 2, \cdots, n$, then

a) $[ww...w] = 0$;
b) $[0\,w_1...w_{n-1}] = 0$;
c) $[w_1 w_2 ...0] = 0$;
d) $[w_1 w_2 ... w_n] = 0$ if any of vectors $w_i$ is the null vector.

(C2) The position change among two vectors in the product
$W = [w_1 w_2 \cdots w_n]$ results in the vector -W.

(C3) If $w_i$ is any vector in space $IR^n$ for $i = 1, 2, \cdots, n$, and $a \in IR$, then

a) $[(aw_1)w_2...w_n] = [w_1(aw_2)...w_n]$;

b) $[(aw_1)w_2...w_n] = a[w_1 w_2 ...w_n]$.

These and other algebraic properties, including the distributive property of the product in relation to the sum of vectors, are verified easily

by the application of the convenient rules on determinants to the matrix structure of product of vectors.

The analogies developed appear still for the possibility of new extensions associated to the concept of products of vectors, such as eventual developments that are related to a type of equivalent n-dimensional of the concept of curl, for example.

## REFERENCES

1. B. Eckmann, "Stetige Lösungen Linearer Gleichungssysteme," Commentarii Mathematici Helvetici, Vol. 15, 1943, pp. 318-339. doi:10.1007/BF02565648

2. N. Efimov, "Elementos de Geometria Analítica," Cultura Brasileira, São Paulo, 1972.

3. A. Elduque, "Vector Cross Products," Talk Presented at the Seminario Rubio de Francia of the Universidad de Zaragoza on April 1 2004.

4. S. Lipschutz and M. Lipson, "Álgebra Linear," Bookman, Porto Alegre, 2008.

5. R. Brown and A. Gray, "Vector Cross Products," Commentarii Mathematici Helvetici, Vol. 42, 1967, pp. 222- 236. doi:10.1007/BF02564418

6. A. Gray, "Vector Cross Products on Manifolds," University of Maryland, College Park, 1968.

7. P. Gritzmann and V. Klee, "On the Complexity of Some Basic Problems in Computational Convexity II. Volume and Mixed Volumes," In: T. Bisztriczky, P. Mc-Muffen, R. Schneider and A. W. Weiss, Eds., Polytopes: Abstract, Convex and Computational, Kluwer, Dordrecht, 1994, p. 29.

8. D. M. Y. Sommerville, "An Introduction to the Geometry of n Dimensions," Dover, New York, 1958, p. 124.

## CITATION

L. Simal Moreira, "Geometric Analogy and Products of Vectors in n Dimensions," Advances in Linear Algebra & Matrix Theory, Vol. 3 No. 1, 2013, pp. 1-6. doi: 10.4236/alamt.2013.31001.

# Generalized Krein Parameters of a Strongly Regular Graph

## Luís Almeida Vieira[1] and Vasco Moço Mano[2]

[1]CMUP—Center of Research of Mathematics of University of Porto, Department of Civil Engineering, University of Porto, Porto, Portugal
[2]CIDMA—Center for Research and Development in Mathematics and Applications

## ABSTRACT

We consider the real three-dimensional Euclidean Jordan algebra associated to a strongly regular graph. Then, the Krein parameters of a strongly regular graph are generalized and some generalized Krein admissibility conditions are deduced. Furthermore, we establish some relations between the classical Krein parameters and the generalized Krein parameters.

## INTRODUCTION

In this paper we explore the close and interesting relationship of a three-dimensional Euclidean Jordan algebra V to the adjacency matrix of a strongly regular graph X. According to [1], the Jordan algebras were formally introduced in 1934 by Pascual Jordan, John von Neumann and Eugene Wigner in [2]. There, the authors attempted to deduce some of the Hermitian matrix properties and they came across a structure lately called a Jordan algebra. Euclidean Jordan algebras were born by adding an inner product with a certain property to a Jordan algebra. It is remarkable that Euclidean Jordan algebras turned out to have such a wide range of applications. For instance, we may cite the application of this theory to statistics [3], interior point methods [4] [5] and combinatorics [6]. More detailed literature on Euclidean Jordan algebras can be found in Koecher's lecture notes [7] and in the monograph by Faraut and Korányi [8].

Along this paper, we consider only simple graphs, i.e., graphs without loops and parallel edges, herein called graphs. Considering a graph X, we denote its vertex set by V(X) and its edge set by E(X)—an edge whose endpoints are the vertices x and y is denoted by xy. In such case, the vertices x and y are adjacent or neighbors. The number of vertices of X, |V(X)|, is called the order of X.

A graph in which all pairs of vertices are adjacent (non-adjacent) is called a complete (null) graph. The number of neighbors of a vertex v in V(X) is called the degree of v. If all vertices of a graph X have degree k, for some natural number k, then X is k-regular.

We associate to X an n by n matrix $A= [a_{ij}]$, where each $a_{ij}=1$, if $v_iv_j \in E(X)$, otherwise $a_{ij}=0$, called the adjacency matrix of X. The eigenvalues of A are simply called the eigenvalues of X.

A non-null and not complete graph X is (n, k, a, c)-strongly regular; if it is k-regular, each pair of adjacent vertices has a common neighbors and each pair of non-adjacent vertices has c common neighbors. The parameters of a (n, k, a, c)-strongly regular graph are not independent and are related by the equality

$$k(k-a-1) = (n-k-1)c.$$

(1)

It is also well known (see, for instance, [9]) that the eigenvalues of a (n, k, a, c)-strongly regular graph X are k, $\theta$ and $\tau$, where $\theta$ and $\tau$ are given by

$$\theta = \left( a - c + \sqrt{(a-c)^2 + 4(k-c)} \right) \Big/ 2,$$

(2)

$$\tau = \left( a - c - \sqrt{(a-c)^2 + 4(k-c)} \right) \Big/ 2.$$

(3)

Therefore, the usually called restricted eigenvalues $\theta$ and $\tau$ are such that the former is positive and the latter is negative. Their multiplicities can be obtained as follows (see, for instance, [10] ):

$$\mu_1 = \frac{1}{2}\left( n - 1 - \frac{(\theta+\tau)(n-1)+2k}{\theta-\tau} \right),$$

(4)

$$\mu_2 = \frac{1}{2}\left( n - 1 + \frac{(\theta+\tau)(n-1)+2k}{\theta-\tau} \right).$$

(5)

Taking into account the above eigenvalues and their multiplicities, the following additional conditions are widely used as feasible conditions for parameters sets (n, k, a, c) of strongly regular graphs; that is, if (n, k, a, c) is a parameter set of a strongly regular graph, then the equality (1) and each one of the following inequalities holds:

- The nontrivial Krein conditions obtained in [11]:

$$(\theta+1)(k+\theta+2\theta\tau) \le (k+\theta)(\tau+1)^2,$$

(6)

$$(\tau+1)(k+\tau+2\theta\tau) \le (k+\tau)(\theta+1)^2.$$

(7)

- The Seidel's absolute bounds qre (see [12] ):

$$n \leq \frac{\mu_1(\mu_1 + 3)}{2} \quad \text{and} \quad n \leq \frac{\mu_2(\mu_2 + 3)}{2}.$$

(8)

With these conditions, many of the parameter sets are discarded as possible parameters sets of strongly regular graphs. To decide whether a set of parameters is the parameter set of a strongly regular graph is one of the main problems on the study of strongly regular graphs. It is worth noticing that these Krein conditions and the Seidel's absolute bounds are special cases of general inequalities obtained for association schemes.

An association scheme with d classes is a finite set S together with d+1 relations $R_i$ defined on S satisfying the following conditions.

1. The set of relations $\{R_0, R_1, \ldots, R_d\}$ is a partition of the Cartesian product of S×S.

$$R_0 = \{(x,x) : x \in S\}.$$

2. If $(x,y) \in R_i$, then also $(y,x) \in R_i$, $\forall x, y \in S$ and for i=0,...,d

3. For each $(x,y) \in R_k$, the number $P_{ij}^k$ of elements $z \in S$ such that $(x,z) \in R_i$ and $(z,y) \in R_j$ depends only from i, j and k.

The numbers $P_{ij}^k$ are called the intersection numbers of the association scheme. Some authors call this type of association schemes symmetric association schemes. The relations $R_i$ of the association scheme can be re-presented by their adjacency matrices $A_i$ of order n=|S| defined by

$$\left(A_i\right)_{xy} = \begin{cases} 1, & \text{if } (x,y) \in R_i, \\ 0, & \text{otherwise.} \end{cases}$$

We may say that $A_i$ is the adjacency matrix of the graph $G_i$, with $V(G_i)=S$ and $E(G_i)=R$. The Bose-Mesner algebra of the association scheme (introduced in [13]) is defined, using these matrices, by the following conditions, which are equivalent to the conditions 1) - 4) of the association scheme:

1. $\sum_{i=0}^{d} A_i = J_n$

2. $A_0 = I_n$,

3. $A_i = A_i^{\mathrm{T}}, \quad \forall i \in \{0, \cdots, d\}$,

4. $A_i A_j = \sum_{k=0}^{d} p_{ij}^{k} A_k, \quad \forall i, j \in \{0, \cdots, d\}$,

where $J_n$ is the matrix of order n whose entries are equal to one and $I_n$ is the identity matrix of order n. From 1) we may conclude that the matrices $A_i$ are linearly independent, and from 2) - 4) it follows that they generate a commutative (d+1)-dimensional algebra A of symmetric matrices with constant diagonal. The matrices $A_i$ commute and then, they can be diagonalized simultaneously, i.e., there exists a matrix B such that $\forall A \in A$, $B^{-1}AB$ is a diagonal matrix. Thus, the algebra A is semisimple and has a unique complete system of orthogonal idempotents $E_0, \ldots, E_d$. Therefore, $\sum_{i=0}^{d} E_i = I_n$ and $E_i E_j = \delta_{ij} E_i$, where

$$\delta_{ij} = \begin{cases} 1, & \text{if } i = j, \\ 0, & \text{otherwise.} \end{cases}$$

This paper is organized as follows. In Section 2, a short introduction on Euclidean Jordan algebras with the fundamental concepts is presented. In order to obtain new feasible conditions for the existence of a strongly regular graph, in Section 3, we define the generalized Krein parameters of a strongly regular graph. In Section 4, we establish some relations between the Krein parameters and the generalized Krein parameters, and present some properties of the generalized Krein parameters. Finally, since the generalized Krein parameters are nonnegative we establish new admissibility conditions, for the parameters of a strongly regular graph that give different information from that given by the Krein conditions 6) - 7).

## EUCLIDEAN JORDAN ALGEBRAS AND STRONGLY REGULAR GRAPHS

In this section the main concepts of Euclidean Jordan Algebras that can be seen for instance in [8], are shortly surveyed.

Let V be a real vector space with finite dimension and a bilinear mapping $(u, v) \mapsto u \circ v$ from V×V to V, that satisfies $(u \circ u) \circ u = u \circ (u \circ u)$, $\forall u \in V$. Then, V is called a real power associative algebra. If V contains an element, e, such that for all u in V, $e \circ u = u \circ e = u$, then e is called the unit element of V. Considering a bilinear mapping $(u, v) \mapsto u \circ v$, if for all u and v in V we have $(J_1)$] $u \circ v = v \circ u$ and $(J_2)$ $u \circ (u^2 \circ v) = u^2 \circ (u \circ v)$, with $u^2 = u \circ u$, then V is called a Jordan algebra. If V is a Jordan algebra with unit element, then V is power associative (cf. [8]). Given a Jordan algebra V with unit element e, if there is an inner product $\langle .. \rangle$ that verifies the equality $\langle u \circ v, w \rangle = \langle v, u \circ w \rangle$, for any u, v, w in V, then V is called an Euclidean Jordan algebra. An element c in an Euclidean Jordan algebra V, with unit element e, is an idempotent if $c^2 = c$. Two idempotents c and d are orthogonal if $c \circ d = 0$. We call the set {$c_1$, $c_2$,...,$c_k$} a complete system of orthogonal idempotents if

I.  $c_i^2 = c_i$, $\forall i \in \{1,...,k\}$;

II.  $c_i \circ c_j = 0$, $\forall i \neq j$ and

III.  $c_1 + c_2 + ... + c_k = e$

Let V be an Euclidean Jordan algebra with unit element e. Then, for every u in V, there are unique distinct real numbers $\lambda_1, \lambda_2, ..., \lambda_k$, and an unique complete system of orthogonal idempotents $\{c_1, c_2, ..., c_k\}$ such that

$$u = \lambda_1 c_1 + \lambda_2 c_2 + \cdots + \lambda_k c_k,$$

(9)

with $c_j \in \mathbb{R}[u]$, j=1,...,k (see [8], Theorem III 1.1). These $\lambda_j$'s are the eigenvalues of u and (9) is called the first spectral decomposition of u.

The rank of an element u in V is the least natural number k, such that the set $\{e, u, ..., u^k\}$ is linear dependent (where $u^k = u \circ u^{k-1}$), and we write rank($\mu$)=k. This concept is expanded by defining the rank of the algebra V as the natural number rank (V) = max {rank (u):u $\in$ V}. The elements of V with rank equal to the rank of V are the regular elements of V. This set of regular elements is open and dense in V. If u is a regular element of V, with r =rank (u), then the set $\{e, u, u^2, ..., u^r\}$ is linearly dependent and the set $\{e, u, u^2, ..., u^{r-1}\}$ is linearly independent. Thus we may conclude that there exist unique real numbers $a_1(u), ..., a_r(u)$, such that $u^r - a_1(u)u^{r-1} + ... + (-1)^r a_r(u)e = 0$), where 0 is the null vector of V. Therefore, with the necessary adjustments, we obtain the following polynomial in $\lambda$ : $p(u, \lambda) = \lambda^r - a_1(u)\lambda^{r-1} + ... + (-1)^r a_r(u)$. This polynomial is called the characteristic polynomial of u, where each coefficient $a_i$ is a homogeneous polynomial of degree i in the coordinates of u in a fixed basis of V. Although we defined the characteristic polynomial for a regular element of V, we can extend this definition to all the elements in V, because each polynomial $a_i$ is homogeneous and, as above referred, the set of regular elements of V is dense in V. The roots of the characteristic polynomial of u, $\lambda_1, \lambda_2, ..., \lambda_r$ are called the eigenvalues of u. Furthermore, the coefficients $a_1(u)$ and $a_r(u)$ in the characteristic polynomial of u, are called the trace and the determinant of u, respectively.

From now on, we consider the Euclidean Jordan algebra of real symmetric matrices of order n, V, such that $\forall A, B \in V$, $A \circ B = (AB+BA)/2$, where AB is the usual product of matrices. Furthermore, the inner product of V is defined as, where tr is the classical trace of matrices, that is the sum of its eigenvalues.

Let X be a (n, k, a, c)-strongly regular graph such that $0<c<k<n-1$, and let A be the adjacency matrix of X. Then A has three distinct eigenvalues, namely the degree of regularity k, and the restricted eigenvalues $\theta$ and $\tau$, given in (2) and (3). Now we consider the Euclidean Jordan subalgebra of V, $V'$, spanned by the identity matrix of order n, $I_n$, and the powers of A. Since A has three distinct eigenvalues, then $V'$ is a three dimensional Euclidean Jordan algebra with $\text{rank}(V') = 3$ and B= $\{I_n, A, A^2\}$ is a basis of $V'$.

Let S= $\{E_0, E_1, E_2\}$ be the unique complete system of orthogonal idempotents of $V'$ associated to A. Then

$$E_0 = \frac{A^2 - (\theta + \tau)A + \theta\tau I_n}{(k-\theta)(k-\tau)} = \frac{J_n}{n},$$

$$E_1 = \frac{A^2 - (k+\tau)A + k\tau I_n}{(\theta-\tau)(\theta-k)},$$

$$E_2 = \frac{A^2 - (k+\theta)A + k\theta I_n}{(\tau-\theta)(\tau-k)},$$

(10)

where $J_n$ is the matrix whose entries are all equal to 1. Since $V'$ is an Euclidean Jordan algebra that is closed for the Hadamard product of matrices, denoted by • and S is a basis of $V'$, then there exist real numbers $q_{\alpha 2}^p$ and $q_{\alpha\beta 11}^p$, $1 \leq \alpha$, $\beta \leq 3$, $\alpha \neq \beta$, such that

$$E_\alpha \bullet E_\alpha = \sum_{p=0}^{2} q_{\alpha 2}^p E_p, \quad E_\alpha \bullet E_\beta = \sum_{p=0}^{2} q_{\alpha\beta 11}^p E_p.$$

$$(11)$$

The real numbers, defined in (11), (whose notation will be clarified later) $q_{\alpha 2}^p$ and $q_{\alpha\beta 11}^p$, $1 \le \alpha$, $\beta \le 3$, $\alpha \ne \beta$, are called the "classical" Krein parameters of the graph X (cf. [10] ). Since $q_{12}^1 \ge 0$ and $q_{22}^2 \ge 0$, the "classical" Krein admissibility conditions $\theta\tau^2 - 2\theta^2\tau - \theta^2 - k\theta + k\tau^2 + 2k\tau \ge 0$, and $\theta^2\tau - 2\theta^2\tau - \tau^2 - k\tau + k\theta^2 + 2k\theta \ge 0$ (presented in [9], Theorem 21.3) can be deduced.

## A GENERALIZATION OF THE KREIN PARAMETERS

Herein the generalized Krein parameters of a (n, k, a, c)-strongly regular graph are defined and then, necessary conditions for the existence of a (n, k, a, c)-strongly regular graph are deduced. These conditions are generalizations of the Krein conditions (see Theorem 21.3 in [9]). Throughout this paper we use a slight different notation from classical books like [9] [14], because, in this way, the connections between the "classical" and the generalized parameters are better understood. Now we generalize the Krein parameters in order to obtain new generalized admissibility conditions on the parameters of strongly regular graphs. Firstly, considering S= {$E_0$, $E_1$, $E_2$} defined like in (10) in the Basis B, and rewriting the idempotents under the new basis {$I_n$, A, $J_n$-A-$1_n$} of $V'$ we obtain

$$E_0 = \frac{\theta - \tau}{n(\theta - \tau)} I_n + \frac{\theta - \tau}{n(\theta - \tau)} A + \frac{\theta - \tau}{n(\theta - \tau)} (J_n - A - I_n),$$

$$E_1 = \frac{|\tau| n + \tau - k}{n(\theta - \tau)} I_n + \frac{n + \tau - k}{n(\theta - \tau)} A + \frac{\tau - k}{n(\theta - \tau)} (J_n - A - I_n),$$

$$E_2 = \frac{\theta n + k - \theta}{n(\theta - \tau)} I_n + \frac{-n + k - \theta}{n(\theta - \tau)} A + \frac{k - \theta}{n(\theta - \tau)} (J_n - A - I_n).$$

$$(12)$$

Consider the natural number p and denote by $M_n(\mathbb{R})$ the set of square matrices of order n with real entries. Then for $B \in M_n(\mathbb{R})$, we denote by $B^{\bullet P}$ and $B^{\otimes P}$ the Hadamard power of order P of B and the Kronecker power of order P of B, respectively, with $B^{\bullet 1} = B$ and $B^{\otimes 1} = B$.

Now, we introduce the following compact notation for the Hadamard and the Kronecker powers of the elements of S. Let x, y, z, α, βand γ be natural numbers such that $1 \le \alpha, \beta, \gamma \le 3$, $x \ge 2$ and $\alpha < \beta$. Then we define

$$E_\alpha^x = (E_\alpha)^x \text{ and } E_\alpha^{\otimes x} = (E_\alpha)^{\otimes x},$$

$$E_{\alpha\beta}^{yz} = (E_\alpha)^y \bullet (E_\beta)^z \text{ and } E_{\alpha\beta}^{\otimes yz} = (E_\alpha)^{\otimes y} \otimes (E_\beta)^{\otimes z},$$

Again, since the Euclidean Jordan algebra $V'$ is closed under the Hadamard product and S is a basis of $V'$, then there exist real numbers $q_{\alpha x}^i$, $q_{\alpha\beta yz}^i$, $q_{(a \oplus \beta)yz}^i$ and $q_{y(a \oplus \beta)yz}^i$, such that

$$E_\alpha^x = \sum_{i=0}^{2} q_{\alpha x}^i E_i, E_{\alpha\beta}^{yz} = \sum_{i=0}^{2} q_{\alpha\beta yz}^i E_i$$

$$(13)$$

We call the parameters $q_{\alpha x}^i$ and $q_{\alpha\beta yz}^i$ defined in (13) the generalized Krein parameters of the strongly regular graph X. Notice that $q_{\alpha 2}^i$ and $q_{\alpha\beta 11}^i$ are precisely the Krein parameters of X already presented. With this notation, the Greek letters are used as idempotent indices and the Latin letters are used as exponents of Hadamard (Kronecker) powers.

## RELATIONS BETWEEN THE KREIN PARAMETERS AND THE GENERALIZED KREIN PARAMETERS

In this section we prove that the generalized Krein parameters can be expressed in function of the Krein parameters. Before that, it is worth to mention that the previously introduced generalizations are straightforward extended to the Krein parameters of symmetric association schemes with ($d \geq 3$) d classes, see [9]. Notice that the algebra spanned by the matrices of a symmetric association scheme with d classes is an Euclidean Jordan Algebra with rank d+1 and with the Jordan product $A \circ B = \dfrac{AB + BA}{2}$ where AB is the usual product of matrices. Furthermore, the inner product of V is defined as $\langle A, B \rangle = \mathrm{tr}(AB)$ where tr(.) is the classical trace of matrices, that is, the sum of its eigenvalues. Let us consider the matrices P and Q of the Bose-Mesner algebra of an association scheme with d classes as defined in [14]. However, for convenience, we denote this matrix Q such as defined in [14] by Q*. Therefore, we can say that $Q = \dfrac{1}{n} Q*$, see [14]. Hence, we can say that the matrices P and Q satisfy,

$$Q_{ij} Q_{ik} = \sum_{l=0}^{2} q_{jk}^{l} Q_{il} \tag{14}$$

$$\left| Q_{ij} \right| \leq \frac{\mu_j}{n} \tag{15}$$

$$\left| P_{ij} \right| \leq n_j \tag{16}$$

$$\sum_{i=0}^{2} n_i Q_{ij} Q_{ik} \leq \frac{u_j}{n} \delta(j,k) \tag{17}$$

The matrices P and Q are usually called the eigenmatrix and the dual eigenmatrix of the association scheme, respectively.

**Theorem 1:** Let G be a (n, k, a, c)-strongly regular graph such that $0<c\leq k<n-1$ whose adjacency matrix is A and has the eigenvalues k, $\theta$ and $\tau$ and whose eigenmatrix and dual eigenmatrix matrix are respectively P and Q If j, k and I are natural numbers such that $0<j$, k, $I\leq2$, then

$$q^l_{jk11} = \sum_{i=0}^{2} Q_{ij} Q_{ik} P_{li}.$$

(18)

**Proof:** Consider that $\{E_0, E_1, E_2\}$ is the of idempotents defined in (12) and the following notation $A_0 = I_n$, $A_1 = A$ and $A_2 = J_n - A - I_n$.

For j, k, $I \in \{0,1,2\}$ since $E_j = \sum_{i=0}^{2} Q_{ij} A_i$ and $E_k = \sum_{i=0}^{2} Q_{ik} A_i$, it follows that $E_j E_k = \sum_{i=0}^{2} Q_{ij} Q_{ik} A_i$

Therefore $E_j E_k E_l = \sum_{i=0}^{2} Q_{ij} Q_{ik} A_i E_l$ and since $A_j = \sum_{t=0}^{2} P_{tj} E_t$ implies $A_i E_l = P_{li} E_l$ we obtain

$$E_j E_k E_l = \sum_{i=0}^{2} Q_{ij} Q_{ik} Q_{li} E_l$$

(19)

Finally, from (19) the result follows.

**Theorem 2:** Let G be a (n, k, a, c)-strongly regular graph such that $0<c\leq k<n-1$ whose adjacency matrix is A and has the eigenvalues k, $\theta$ and $\tau$ and whose eigenmatrix and dual eigenmatrix matrix are respec-

tively P and Q Let j, m and s be natural numbers such that $0 \le j, s \le 2$. Then

$$q^{s}_{jm} = \sum_{i=0}^{2} \left( Q_{ij} \right)^{m} P_{si}.$$

(20)

**Proof:** Taking into account that $Q^{m}_{j} = \sum_{i=0}^{2} q^{i}_{jm} E_{i}$ and by the equalities (21) and (22)

$$E_{j} = \sum_{i=0}^{2} Q_{ij} A_{i},$$

(21)

$$A_{i} = \sum_{j=0}^{2} P_{ji} E_{j},$$

(22)

we conclude that $E^{m}_{j} = \sum_{i=0}^{2} (Q_{ij})^{m} A_{i}$ Therefore $q^{s}_{jm} E_{s} = E^{m}_{j} E_{s} = \sum_{i=0}^{2} (Q_{ij})^{m} A_{i} E_{s}$ and since by (22)

$A_{i} E_{s} = P_{si} E_{s}$ we obtain. Hence $q^{s}_{jm} E_{s} = \sum_{i=0}^{2} (Q_{ij})^{m} P_{si} E_{s}$

As an application of the Theorem 2 we may conclude that considering a strongly regular graph G the generalized Krein parameters $q^{l}_{jm}$ can be expressed in function of the classical Krein parameters as follows:

$$q^{s}_{jm} = \sum_{l_{1}=0 l_{2}=0}^{2} \sum_{l_{n-2}=0}^{2} \cdots \sum_{l_{n-2}=0}^{2} q^{l_{1}}_{j2} q^{l_{2}}_{l_{1}j11} \cdots q^{l_{m-2}}_{(m-3)j11} q^{s}_{l_{m-2}j11}.$$

(23)

The expression (23) is obtained using (14) and (20). Summarizing, we have the following corollary.

Corollary 1. Let G be a (n, k, a, c)-strongly regular graph such that $0<c\leq k<n-1$ Then for all natural numbers j,m and s such that

$$q^s_{j(2+m)} = \sum_{l_1=0}^{2}\sum_{l_2=0}^{2}\cdots\sum_{l_m=0}^{2} q^{l_1}_{j2}q^{l_2}_{l_1 j11}\cdots q^{l_m}_{l_{m-1}j11}q^s_{l_m j11}.$$

(24)

**Theorem 3:** Let G be a (n, k, a, c)-strongly regular graph such that $0<c\leq k<n-1$ Then for all natural numbers i, j, m, n, and s such that $0\leq i,j,s\leq 2$,

$$q^s_{ijmn} = \sum_{l=0}^{2}(Q_{li})^m (Q_{lj})^n P_{sl}.$$

(25)

**Proof:** We have $E^m_i E^n_j = \sum_{l=0}^{2} q^l_{ijmn}E_l$ Since from (21) $E^n_i E^m_j = \sum_{l=0}^{2}(Q_{li})^n$ $A_l \sum_{l=0}^{2}(Q_{li})^m A_l$ then

$E^m_i E^n_j = \sum_{l=0}^{2} q^l_{ijmn}E_l$. Hence we obtain

$E^n_i E^m_j = \sum_{l=0}^{2}(Q_{li})^n (Q_{lj})^m P_{sl}E_s$

Therefore, the equality (25) follows. ,

Recurring to (14) and (25), we may conclude the Corollary 2.

*Corollary 2:* Let G be a $(n, k, a, c)$-strongly regular graph such that $0 < c \leq k < n-1$ Then for all natural numbers $i_1$, $i_2$, $m$, $n$ and $s$ such that $0 \leq i_1$, $i_2$, $s \leq 2$,

$$q^s_{i_1 i_2 nm} = \sum_{i_1=0}^{2} \cdots \sum_{i_n=0}^{2} \sum_{i_{n+1}=0}^{2} \cdots \sum_{i_{n+m-2}=0}^{2} q^{i_1}_{i_1 2} q^{i_2}_{i_1 i_1 11} \cdots q^{i_{n-1}}_{i_{n-2} i_1 11} q^{i_n}_{i_{n-1} i_2 11} q^{i_{n+1}}_{i_n i_2 11} \cdots q^{i_{n+m-2}}_{i_{n+m-3} i_2 11} q^{s}_{i_{n+m-2} 2 i_2 11}.$$

*Theorem 4:* Let G be a $(n, k, a, c)$-strongly regular graph such that $0 < c \leq k < n-1$ Then $\forall n \in \mathbb{N}$ and $\forall i_1, \ldots, i_{n+1} \in \{0, 1, 2\}$,

$$\sum_{r=0}^{2} Q_{r i_1} \cdots Q_{r i_{n+1}} P_{sr} \geq 0.$$
(26)

*Proof:* We prove by induction on n. For $n=1$ the inequality (26) holds, since the classical Krein parameters $q^s_{i_1 i_2 11}$ are nonnegative and $\sum_{r=0}^{2} Q_{r i_1} Q_{r i_2} P_{sr} = q^s_{i_1 i_2 11}$ Now assuming that the inequality (26) holds for $n = k \geq 1$, we prove that (26) also holds for $n = k+1$ Consider the sum $\sum_{r=0}^{2} Q_{r i_1} \cdots Q_{r i_{k+2}} P_{sr}$ Then from (14), we obtain

$$\sum_{r=0}^{2} Q_{r i_1} \cdots Q_{r i_{k+2}} P_{sr} = \sum_{r=0}^{2} \sum_{l=0}^{2} q^l_{i_1 i_2 11} Q_{rl} Q_{r i_3} \cdots Q_{r i_{k+2}} P_{sr}$$

$$= \sum_{l=0}^{2} q^l_{i_1 i_2 11} \sum_{r=0}^{2} Q_{rl} Q_{r i_3} \cdots Q_{r i_{k+2}} P_{sr}.$$

Since for $n=k$ the inequality (26) is verified and the summands $\sum_{r=0}^{2} Q_{rl} Q_{r i_3} \cdots Q_{r i_{k+2}} P_{sr}$ are nonnegative, we may conclude that $\sum_{r=0}^{2} Q_{r i_1} \cdots Q_{r i_{n+1}} P_{sr} \geq 0$ ,

Recurring to the Theorem 4 we are conducted to the Corollaries 3 and 4.

Corollary 3. Let G be a (n, k, a, c)-strongly regular graph such that $0<c\leq k<n-1$ Then for all natural numbers i, m, s and s such that $0\leq i, s\leq 2$ the generalized Krein parameters $q^s_{im}$ are nonnegative.

**Corollary 4:** Let G be a (n, k, a, c)-strongly regular graph such that $0<c\leq k<n-1$ Then for all natural numbers I, j, m, n and s such that $0\leq i, j, s\leq 2$ the generalized Krein parameters $q^s_{ijmn}$ are nonnegative.

**Theorem 5:** Let G be a strongly regular graph and let i, s and m be natural numbers such that $0\leq I, s\leq 2$ Then $q^s_{im} \leq 1$

**Proof:** Recurring to the inequalities (14)-(17) we have:

$$q^s_{im} = \sum_{t=0}^{2}\left(Q(t,i)\right)^m P(s,t) = \left|\sum_{t=0}^{2}\left(Q(t,i)\right)^{m-2}\left(Q(t,i)\right)^2 P(s,t)\right|$$

$$\leq \sum_{t=0}^{2}\left|\left(Q(t,i)\right)^{m-2}\right|\left|\left(Q(t,i)\right)^2\right|\left|P(s,t)\right| \leq \sum_{t=0}^{2}\left(\frac{\mu_i}{n}\right)^{m-2}\left(Q(t,i)\right)^2 n_t$$

$$\leq \sum_{t=0}^{2}1^{m-2}\left(Q(t,i)\right)^2 n_t \leq \frac{\mu_i}{n} \leq 1.$$

**Theorem 6:** Let G be a (n, k, a, c)-strongly regular graph such that $0<c\leq k<n-1$. Let i, j, m, n and s be natural numbers such that $0\leq i, j, s\leq 2$ and $m+n\geq 3$ Then the generalized Krein parameter $q^s_{ijmn}$ satisfy $q^s_{ijmn} \leq 1$

**Proof:** Similar to the Proof done in Theorem 5. ꙩ

Let G be a (n, k, a, c)-strongly regular graph such that $0<c\leq k<n-1$ since the generalized Krein parameters $q^s_{i_1i_2mn}$ and $q^s_{i_1mn}$ are nonnegative

then we can establish new admissibility conditions distinct from the Krein conditions (6) and (7). For instance, the generalized Krein condition $q^0_{113} \geq 0$ allows us to establish a new theorem on strongly regular graphs after some algebraic manipulation of its expressions. Analyzing the generalized Krein parameter $q^0_{113}$ of a strongly regular graph with one in its spectra we deduce the following theorem (7).

**Theorem 7:** Let G be a (n, k, a, c)-strongly regular graph such that $0 < c \leq k < n-1$ whose adjacency matrix is A and has the eigenvalues k, $\theta$ =1 and $\tau$ If $k \geq 9$ then

$$n \leq \frac{3}{2} + \frac{53}{20} k.$$

(27)

**Proof:** Since $q^0_{113} \geq 0$ then we have

$$\left( \frac{\theta n + k - \theta}{n(\theta - \tau)} \right)^3 + \left( \frac{-n + k - \theta}{n(\theta - \tau)} \right)^3 k + \left( \frac{k - \theta}{n(\theta - \tau)} \right)^3 (n - k - 1) \geq 0.$$

(28)

From the inequality (28) and after some simplifications we conclude that

$$n^2 \left( \theta^3 - k \right) + n \left( -3\theta^3 + 3k^2 + 3\theta^2 k - 3k\theta \right) + \left( 2\theta^3 - 2k^3 + 6\theta k^2 - 6\theta^2 k \right) \geq 0.$$

Therefore if $\theta = 1$ then

$$n^2 \left( 1 - k \right) + n \left( -3 + 3k^2 + 3k - 3k \right) + \left( 2 - 2k^3 + 6k^2 - 6k \right) \geq 0$$

Finally we have

$$n^2\left(1-k\right)+n\left(-3+3k^2\right)+2\left(1-k^3+3k^2-3k\right)\ge 0$$

(29)

Dividing both members of (29) by 1-k we are supposing that k>1 we obtain

$$n^2-3n\left(1+k\right)+2\left(k^2-2k+1\right)\le 0.$$

(30)

[1]We must note that the equation $x^2-3x(1+k)+2(1-2k+k^2)=0$ as the

roots $x_2=\dfrac{3(k+1)-\sqrt{k^2-34k+1}}{2}$ and the root $x_2=\dfrac{3(k+1)+\sqrt{k^2+34k+1}}{2}$

since k>8 implies that k²+34k+1≤5k² and finally this implies that

$$x_2\le\dfrac{3(k+1)+\sqrt{5k}}{2}\text{ therefore }x_2\le\dfrac{3}{2}+\dfrac{53k}{20}$$

Now from the inequality (30) we conclude that if G is a (n, p, a, c)-strongly regular graph with one in his

Spectra1 then k≥9 implies that $n\le\dfrac{3}{2}+\dfrac{53}{20}k$

We now present in Table 1 some examples of parameter sets (n, k, a, c) that do not verify the inequality (27) of Theorem 7. We consider the parameter sets $P_1(28,9,0,4)$, $P_2(64,21,0,10)$, $P_3=(1225,456,39,247)$, $P_4=(1296,481.0,40.0,260)$ and $P_5=(1024,385,36,210)$. For each example we present the respective eigen-Values $\theta$, $\tau$ and the value of $q_{kn}$

defined by $q_{kn}=\dfrac{53}{20}k+\dfrac{3}{2}-n$

## SOME CONCLUSIONS

In this paper, we have generalized the Krein parameters of a strongly regular graph and obtained some relations between the classical Krein parameters and the generalized Krein parameters (see Corollaries 1 and 2). We also establish that these generalize Krein parameters are always positive and less than one (see Corollaries 3 and 4, and Theorems 5 and 6). Let I, j, m, n and s be natural numbers such that $0 \leq i,\ j, s \leq 2$ The generalized Krein admissibility conditions $q_{ijmn}^s \geq 0$ with $m+n \geq 3$ and $q_{im}^s \geq 0$ with $m \geq 3$ allow us to establish new admissibility conditions; they permit us to establish new inequalities on the parameters of a strongly regular graph. For instance the generalized Krein parameter condition $q_{23}^0 \geq 0$ after some algebraic manipulation allows us to establish the inequality (27) in Theorem 7. Finally, we conclude that we can extend the definition of generalized Krein parameters to a symmetric association scheme with d classes.

**Table 1**: Numerical results when $k \geq 9$

|  | $P_1$ | $P_2$ | $P_3$ | $P_4$ | $P_5$ |
|---|---|---|---|---|---|
| $\theta$ | 1 | 1 | 1 | 1 | 1 |
| $\tau$ | −5 | −11 | −209 | −221 | −175 |
| $q_{\theta\tau kn}^1$ | −2.65 | −6.85 | −15.1 | −19.85 | −2.5 |

## ACKNOWLEDGEMENTS

1. Luís Almeida Vieira was supported by the European Regional Development Fund through the program COMPETE and by the Portuguese Government through the FCT—Fundação para a Ciência e a Tecnologia under the project PEst—C/MAT/UI0144/2013.

2. Vasco Moço Mano was partially supported by Portuguese Funds trough CIDMA—Center for Research and development in Mathematics and Applications, Department of Mathematics, University of Aveiro, 3810-193, Aveiro, Portugal and the Portuguese Foundation for Science and Technology (FCT-Fundação para a Ciência e Tecnologia), within Project PEst-OE/MAT/UI4106/2014.

3. The authors would like to thank the anonymous referee for his/her careful revision and relevant comments that improved our paper.

## REFERENCES

1. McCrimmon, K. (1978) Jordan Algebras and Their Applications. Bulletin of the American Mathematical Society, 84, 612-627. http://dx.doi.org/10.1090/S0002-9904-1978-14503-0

2. Jordan, P., Neuman, J.V. and Wigner, E. (1934) On an Algebraic Generalization of the Quantum Mechanical Formalism. Annals of Mathematics, 35, 29-64. http://dx.doi.org/10.2307/1968117

3. Massan, H. and Neher, E. (1998) Estimation and Testing for Lattice Conditional Independence Models on Euclidean Jordan Algebras. Annals of Statistics, 26, 1051-1082.http://dx.doi.org/10.1214/aos/1024691088

4. Faybusovich, L. (1997) Euclidean Jordan Algebras and Interior-Point Algorithms. Positivity, 1, 331-357. http://dx.doi.org/10.1023/A:1009701824047

5. Faybusovich, L. (2007) Linear Systems in Jordan Algebras and Primal-Dual Interior-Point Algorithms. Journal of Computational and Applied Mathematics, 86, 149-175.http://dx.doi.org/10.1016/S0377-0427(97)00153-2

6. Cardoso, D.M. and Vieira, L.A. (2004) Euclidean Jordan Algebras with Strongly Regular Graphs. Journal of Mathematical Sciences, 120, 881-894.http://dx.doi.org/10.1023/B:JOTH.0000013553.99598.cb

7. Koecher, M. (1999) The Minnesota Notes on Jordan Algebras and Their Applications. Krieg, A. and Walcher, S., Eds., Springer, Berlin.

8. Faraut, J. and Korányi, A. (1994) Analysis on Symmetric Cones. Oxford Science Publications, Oxford.

9. Godsil, C. and Royle, G. (2001) Algebraic Graph Theory. Springer, Berlin.http://dx.doi.org/10.1007/978-1-4613-0163-9

10. van Lint, J.H. and Wilson, R.M. (2004) A Course in Combinatorics. Cambridge University Press, Cambridge.

11. Scott Jr., L.L. (1973) A Condition on Higman's Parameters. Notices of the American Mathematical Society, 20, A-97.

12. Delsarte, Ph., Goethals, J.M. and Seidel, J.J. (1975) Bounds for Systems of Lines and Jacobi Polynomials. Philips Research Reports, 30, 91-105.

13. Bose, R.C. and Mesner, D.M. (1952) On Linear Associative Algebras Corresponding to Association Schemes of Partially Balanced Designs. The Annals of Mathematical Statistics, 47, 151-184.

14. Brower, A.E. and Haemers, W.H. (1995) Association Schemes. In: Grahm, R., Grotsel, M. and Lovász. L., Eds., Handbook of Combinatorics, Elsevier, Amsterdam, 745-771.

## CITATION

Vieira, L. and Mano, V. (2015) Generalized Krein Parameters of a Strongly Regular Graph. Applied Mathematics, 6, 37-45. doi: 10.4236/am.2015.61005.

# A Measurement Theoretical Foundation of Statistics

## Shiro Ishikawa

Department of Mathematics, Faculty of Science and Technology, Keio University, Yokohama, Japan

4

## ABSTRACT

It is a matter of course that Kolmogorov's probability theory is a very useful mathematical tool for the analysis of statistics. However, this fact never means that statistics is based on Kolmogorov's probability theory, since it is not guaranteed that mathematics and our world are connected. In order that mathematics asserts some statements concerning our world, a certain theory (so called "world view") mediates between mathematics and our world. Recently we propose measurement theory (i.e., the theory of the quantum mechanical world view), which is characterized as the linguistic turn of quantum mechanics. In this paper, we assert that statistics is based on measurement theory. And, for example, we show, from the pure theoretical point of view (i.e., from the measurement theoretical point of view), that regression analysis cannot be justified without Bayes' theorem. This may imply that even the conventional classification of (Fisher's) statistics and Bayesian statistics should be reconsidered.

## INTRODUCTION

For example, consider Newtonian mechanics. It is natural to understand that Newton mechanics is based on Newton's three laws of motion, though the mathematical theory of differential equations is a useful tool for the analysis of Newtonian mechanics. That is because any mathematical theory is a closed logical system derived from set theory, and thus, it is not qualified to assert statements concerning our world without laws. If it is so, and, if Kolmogorov's probability theory [1] is a mathematical theory, we think that the foundation of statistics does not yet established. Thus, the following problem is natural:

(A) What kind of law is statistics based on? Or, propose a foundation of statistics!

The purpose of this paper is to answer this problem.

Although in a series of our research [2-8] we have been concerned with this problem (A), in this paper we give a decisive answer to the problem (A) in the light of our final version [7, 8] of measurement theory. Here, as mentioned in Section 2 later, measurement theory (i.e., the theory of the quantum mechanical world view) is characterized as the linguistic turn of quantum mechanics. Hence, note that measurement theory is not physics but a kind of language, and thus, the "law" in (A) is called "axiom" in this paper.

## MEASUREMENT THEORY (AXIOMS AND INTERPRETATION)

### Mathematical Preparations

In this section, we prepare mathematics, which is used in measurement theory (or in short, MT).

Measurement theory ([2-8]) is, by an analogy of quantum mechanics (or, as a linguistic turn of quantum mechanics), constructed as the scientific theory formulated in a certain C*-algebra A (i.e., a norm closed sub algebra in the operator algebra B (H) composed of all bounded operators on a Hilbert space H, cf. [9,10]). MT is composed of two theories (i.e., pure measurement theory (or, in short, PMT] and statistical measurement theory (or, in short, SMT). That is, we see:

(B) MT (measurement theory)

$$
= \begin{cases} (B_1): [PMT] \\ \quad = \underset{(Axiom^P\ 1)}{\big[(pure)\,measurement\big]} + \underset{(Axiom\ 2)}{\big[causality\big]} \\ (B_2): [SMT] \\ \quad = \underset{(Axiom^S\ 1)}{\big[(statistical)\,measurement\big]} + \underset{(Axiom\ 2)}{\big[causality\big]} \end{cases}
$$

Where Axiom 2 is common in PMT and SMT. for completeness, note that measurement theory (B) (i.e., $(B_1)$ and $(B_2)$) is a kind of language based on the quantum mechanical world view, (cf. [8]). It may be understandable to consider that

(C) PMT and SMT is related to Fisher's statistics and Bayesian statistics respectively.

Also, as mentioned in Section 2.6 latter, our concern in this paper is to give an answer to the question "Which is fundamental, PMT or SMT?"

When $A=B_c$ (H), the C*-algebra composed of all compact operators on a Hilbert space H, the (B) is called quantum measurement theory (or, quantum system theory), which can be regarded as the linguistic aspect of quantum mechanics. Also, when A is commutative (that is, when A is characterized by $C_0(\Omega)$, the $C^*$- algebra composed of all continuous complex-valued functions vanishing at infinity on a locally compact Hausdorff space $\Omega$ (cf. [9])), the (B) is called classical measurement theory. Thus, we have the following classification:

$$(D) \quad MT \begin{cases} \text{quantum MT (when } \mathcal{A} = B_c(H)) \\ \text{classical MT (when } \mathcal{A} = C_0(\Omega)) \end{cases}$$

In this paper, we mainly devote ourselves to classical MT (i.e., classical PMT and classical SMT).

Now we shall explain the measurement theory (B). Let $A(\subseteq B(H))$ be a $C^*$-algebra, and let $A^*$ be the dual Banach space of A. That is, $= A^*\{\rho | \rho$ is a continuous linear functional on A$\}$, and the norm $\|\rho\|_{A^*}$ is defined by

$$\sup\left\{ \left|\rho(F)\right| : F \in \mathcal{A} \text{ such that } \|F\|_{\mathcal{A}} \left(= \|F\|_{B(H)}\right) \leq 1 \right\}.$$

The bi-linear functional $\rho(F)$ is also denoted by $_{A^*}\langle \rho, F \rangle_A$, or in short $\langle \rho, F \rangle$. Define the mixed state $\rho(\in A^*)$ such that $\|\rho\|_{A^*} = 1$ and $\rho(F) \geq 0$ for all $F \in A$ satisfying $F \geq 0$. And put

$$\mathfrak{S}^m(\mathcal{A}^*) = \{\rho \in \mathcal{A}^* | \rho \text{ is a mixed state}\}.$$

A mixed state $(\rho \in \mathfrak{S}^m(A^*))$ is called a pure state if it satisfies that $\rho = \theta\rho_1 + (1-\theta)\rho_2$ for some $\rho_1, \rho_2 \in \mathfrak{S}^m(A^*)$ and $0 < \theta < 1$ implies $\rho = \rho_1 = \rho_2$. Put

$$\mathfrak{S}^p(A^*) = \{\rho \in \mathfrak{S}^m(A^*) \mid \rho \text{ is a pure state}\},$$

Which is called a state space. The Riese theorem (cf. [11]) says that

$$C_0(\Omega)^* = M(\Omega) = \{\rho \mid \rho \text{ is a signed measure on } \Omega\},$$

$$\mathfrak{S}^m(C_0(\Omega)^*) = M_{+1}^m(\Omega)$$

$$= \{\rho \mid \rho \text{ is a measure on } \Omega \text{ such that} \rho(\Omega) = 1\}.$$

Also, it is well known (cf. [9]) that

$$\mathfrak{S}^p(B_c(H)^*) = \{|u\rangle\langle u| (i.e., \text{ the Dirac notation}) \mid \|u\|_H = 1\},$$

and

$$\mathfrak{S}^p(C_0(\Omega)^*) = M_{+1}^p(\Omega)$$

$$= \{\delta_{\omega_0} \mid \delta_{\omega_0} \text{ is a point measure at } \omega_0 \in \Omega\},$$

Where $\int_\Omega f(\omega)\delta_{\omega_0}(d\omega) = f(\omega_0)(\forall f \in C_0(\Omega))$. The latter implies that $\mathfrak{S}^p(C_0(\Omega)^*)$ can be also identified with $\Omega$ (called a spectrum space or maximal ideal space) such as

$$\underset{\text{(state space)}}{\mathfrak{S}^p(C_0(\Omega)^*)} \ni \delta_\omega \leftrightarrow \omega \in \underset{\text{(spectrum space)}}{\Omega}$$

Here, assume that the $C^*$-algebra $A(\subseteq B(H))$ is unital, i.e., it has the identity I. This assumption is not unnatural, since, if $I \notin A$, it suffices to reconstruct the A such that it includes $A \cup \{I\}$.

According to the noted idea (cf. [12]) in quantum mechanics, an observable $O \equiv (X, \mathcal{F}, F)$ in A is defined as follows:

($E_1$) [Field] X is a set, $\mathcal{F}(\subseteq 2^X$ the power set of X) is a field of X, that is, "$\Xi_1, \Xi_2 \in \mathcal{F} \Rightarrow \Xi_1 \cup \in \mathcal{F}$", "$\Xi \in \mathcal{F} \Rightarrow X / \Xi \in \mathcal{F}$".

($E_2$) [Countably additivity] F is a mapping from $\mathcal{F}$ to A satisfying:

1. For every $\Xi \in \mathcal{F}$, $F(\Xi)$ is a nonnegative element in A such that $0 \leq F(\Xi) \leq I$,

2. $F(\varnothing) = 0$ and $F(X) = 1$, where 0 and I is the 0- element and the identity in A respectively.

3. For any countable decomposition $\{\Xi_1, \Xi_2, ...\}$ of $\Xi \in \mathcal{F}$ (i.e., $\Xi_k, \Xi \in \mathcal{F}$ such that $\bigcup_{k=1}^{\infty} \Xi_k = \Xi$, $\Xi_i \cap \Xi_j = \varnothing (i \neq j)$, it holds that

$$\lim_{K \to \infty} \rho \left( F\left( \bigcup_{k=1}^{K} \Xi_k \right) \right) = \rho(F(\Xi)) \quad (\forall \rho \in \mathfrak{S}^m(\mathcal{A}^*))$$

(1)

**Remark 1:** By the Hopf extension theorem (cf. [11]), we have the mathematical probability space $(X, \bar{\mathcal{F}}, \rho^m(F(.)))$ where $\bar{\mathcal{F}}$ is the smallest $\sigma$-field such that $F \subseteq \bar{\mathcal{F}}$. For the other formulation (i.e., $W^*$-algebraic formulation), see the appendix in [7].

## Pure Measurement Theory in (B$_1$)

In what follows, we shall explain PMT in (B$_1$).

With any system S, a C*-algebra $A(\subseteq B(H))$ can be associated in which the pure measurement theory (B$_1$) of that system can be formulated. A state of the system S is represented by an element $\rho(\in \mathfrak{S}^m(A^*))$ and an observable is represented by an observable $o = (X, \mathcal{F}, F)$ in A. Also, the measurement of the observable O for the system S with the state $\rho$ is denoted by $M_A\left(o, S_{[\rho]}\right)$ (or more precisely, $M_A\left(0 = (X, \mathcal{F}, F), S_{[\rho]}\right)$ an observer can obtain a measured value $x = (\in X)$ by the measurement $M_A\left(o, S_{[\rho]}\right)$.

The Axiom$^P$ 1 presented below is a kind of mathematical generalization of Born's probabilistic interpretation of quantum mechanics. And thus, it is a statement without reality.

**Axiom$^P$ 1. [Pure Measurement]:** The probability that a measured value $x = (\in X)$ obtained by the measurement $M_A\left(o \equiv (X, \mathcal{F}, F), S_{[\rho_0]}\right)$ belongs to a set $\Xi(\in \mathcal{F})$ is given by $\rho_0\left(F(\Xi)\right)$.

Next, we explain Axiom 2 in (B). Let $(T, \leq)$ be a tree, i.e., a partial ordered set such that $t_1 \leq t_3$ and $t_2 \leq t_3$ implies $t_1 \leq t_2$ or $t_2 \leq t_1$. In this paper, we assume that T is finite (cf. Remark 9 in Section 7 later). Assume that there exists an element $t_0 \in T$, called the root of T, such that $t_0 \leq t$ $((\forall t \in T)$ holds. Put $T_{\leq}^2\{(t_1, t_2) \in T^2 | t_1 \leq t_2\}$. The family $\left\{\Phi_{t_1, t_2} : A_{t_2} \to A_{t_2}\right\}_{(t_1, t_2) \in T_{\leq}^2}$ is called a causal relation (due to the Heisenberg picture), if it satisfies the following conditions (F$_1$) and (F$_2$).

$(F_1)$ With each $t \in T$, a $C^*$-algebra $A_t$ is associated.

$(F_2)$ For every $(t_1, t_2) \in T_\leq^2$, a Markov operator $\Phi_{t_1,t_2} : A_{t_2} \to A_{t_1}$ is defined (i.e., $\Phi_{t_1,t_2} \geq 0, \Phi_{t_1,t_2}\left(1_{A_{t_2}}\right) = 1_{A_{t_1}}$). And it satisfies that $\Phi_{t_1,t_2}\Phi_{t_2,t_3} = \Phi_{t_1,t_3}$ holds for any $(t_1,t_2)(t_1,t_3) \in T_\leq^2$. The family of dual operators $\{\Phi_{t_1,t_2}^* :$ $\mathfrak{S}^m(A_{t_1}^*) \to \mathfrak{S}^m(A_{t_2}^*)\}_{(t_1,t_2)\in T_\leq^2}$ is called a dual causal relation (due to the Schrödinger picture). When $\Phi_{t_1,t_2}(\mathfrak{S}^p(A_{t_1}^*)) \subseteq (\mathfrak{S}^p(A_{t_2}^*))$ holds for an $(t_1,t_2) \in T_\leq^2$, the causal relation is said to be deterministic.

Now Axiom 2 in the measurement theory (B) is presented as follows:

**Axiom 2[Causality]:** The causality is represented by a causal relation $\{\Phi_{t_1,t_2} : A_{t_2} \to A_{t_1}\}_{(t_1,t_2)\in T_\leq^2}$.

## Interpretation

Next, we have to study how to use the above axioms as follows. That is, we present the following interpretation (G) [= $(G_1)$ – $(G_3)$], which is characterized as a kind of linguistic turn of so-called Copenhagen interpretation (cf. [7, 8]). That is, we propose:

1. $(G_1)$ Consider the dualism composed of observer and system (= measuring object). And therefore, observer and system must be absolutely separated.

2. $(G_2)$ Only one measurement is permitted. And thus, the state after a measurement is meaningless since it cannot be measured any longer. Also, the causality should be assumed only in the side of system, however, a state never moves. Thus, the Heisenberg picture

should be adopted, and thus, the Schrödinger picture should be prohibited.

3. $(G_3)$ Also, the observer does not have the space-time. Thus, the question: "When and where is a measured value obtained?" is out of measurement theory. And thus, Schrödinger's cat is out of measurement theory, and so on.

## Sequential Causal Observable and Its Realization

For each $k = 1, 2 \cdots, K$, consider a measurement $M_A\left(O_k \equiv (X_k, \mathcal{F}_k, F_k), S_{[\rho]}\right)$. However, since the $(G_2)$ says that only one measurement is permitted, the measurements $\left\{M_A\left(O_k, S_{[\rho]}\right)\right\}_{k=1}^{K}$ should be reconsidered in what follows. Under the commutatively condition such that

$$F_i\left(\Xi_i\right)F_j\left(\Xi_j\right) = F_j\left(\Xi_j\right)F_i\left(\Xi_i\right)$$
$$\left(\forall \Xi_i \in \mathcal{F}_i, \forall \Xi_j \in \mathcal{F}_j, i \neq j\right),$$

$$(2)$$

We can define the product observable

$$X_{k=1}^{K}O_k = \left(X_{k=1}^{K}X_k \underset{k=1}{\overset{K}{\boxed{\times}}} \mathcal{F}_k, X_{k=1}^{K}F_k\right) \text{ in A such that}$$

$$\left(X_{k=1}^{K}F_k\right)\left(X_{k=1}^{K}\Xi_k\right) = F_1\left(\Xi_1\right)F_2\left(\Xi_2\right)\cdots F_K\left(\Xi_K\right)$$
$$\left(\forall \Xi_k \in \mathcal{F}_k, \forall k = 1, \cdots, K\right).$$

Here, $\underset{k=1}{\overset{K}{\boxed{\times}}} \mathcal{F}_k$ is the smallest field including the family $\left\{X_{k=1}^{K}\Xi_k : \Xi_k \in \mathcal{F}_k k = 1, 2, \cdots, K\right\}$. Then, the above $\left\{M_A\left(O_k, S_{[\rho]}\right)\right\}_{k=1}^{K}$ is, under the commutativity condition (2), represented by the simultaneous measurement $M_A \left( X_{k=1}^K O_k, S_{[\rho]} \right)$.

Consider a tree $\left( T \equiv \{t_0, t_1, \cdots t_n\}, \leq \right)$ with the root $t_0$. This is also characterized by the map $: T \setminus \{t_0\} \circledR T$ such that $\pi(t) = \max \{s \in T \mid s < t\}$. Let $\left\{ \Phi_{t,t'} : A_{t'} \to A_t \right\}_{(t,t') \in T_{\leq}^2}$ be a causal relation, which is also represented by $\left\{ \Phi_{\pi(t),t} : A_t \to A_{\pi(t)} \right\}_{t \in T \setminus \{t_0\}}$. Let an observable $O_t \equiv (X_t, \mathcal{F}_t, F_t)$ in the $A_t$ be given for each $t \in T$. Note that $\Phi_{\pi(t),t} O_t \left( \equiv \left( X_t, \mathcal{F}_t, \Phi_{\pi(t),t} F_t \right) \right)$ is an observable in the $A_{\pi(t)}$.

The pair $[O_T] = \left[ [O_t] \right]_{t \in T} \left\{ \Phi_{t \in t'} : A_{t'} \to A_t \right\}_{(t,t') \in T_{\leq}^2}$, is called a sequential causal observable. For each $s \in T$, put $T_s = \{t \in T \mid t \geq s\}$. And define the observable

$$\hat{O} \equiv \left( \times_{t \in T_s} X_t, \right) \boxtimes_{t \in T_s} \mathcal{F}_t, \hat{F}_s \text{ in } A_s \text{ follows:}$$

$$\hat{O}_s = \begin{cases} O_s & \text{if } s \in T \setminus \pi(T) \\ O_s \times \left( \times_{t \in \pi^{-1}(\{s\})} \Phi_{\pi(t),t} \hat{O}_t \right) & \text{if } s \in \pi(T) \end{cases}$$

(3)

If the commutativity condition holds (i.e., if the product observable $O_s \times (X_{t \in \pi^{-1}(\{s\})} \Phi_{\pi(t),t} \hat{O}_t)$ exists) for each $s \in \pi(T)$. Using (3) iteratively, we can finally obtain the observable $\hat{O}_{t_0}$ in $A_{t0}$. The $\hat{O}_{t_0}$ is called the realization (or, realized causal observable) of $[O_T]$.

**A Measurement Theoretical Foundation of Statistics**

## Statistical Measurement Theory in $(B_2)$

We shall introduce the following notation: it is usual to consider that we do not know the pure state $\rho_0^p (\in \mathfrak{S}^p(A^*))$ when we take a measurement $M_A\left(O, S_{[\rho_0^p]}\right)$. That is because we usually take a measurement $M_A\left(O, S_{[\rho_0^p]}\right)$ in order to know the state $\rho_0^p$. Thus, when we want to emphasize that we do not know the state $\rho_0^p$, $M_A\left(O, S_{[\rho_0^p]}\right)$ is denoted by $M_A\left(O, S_{[*]}\right)$. Also, when we know the distribution $\rho_0^m (\in \mathfrak{S}^m(A^*))$ of the unknown state $\rho_0^p$, the $M_A\left(O, S_{[\rho_0^p]}\right)$ is denoted by $M_A\left(O, S_{[*]}(\{\rho_0^m\})\right)$. The $\rho_0^m$ is called a mixed state. And further, if we know that a mixed state $\rho_0^m$ belongs to a compact set $K(\subseteq \mathfrak{S}^m(A^*))$, the $M_A\left(O, S_{[\rho_0^p]}\right)$ is denoted by $M_A\left(O, S_{[*]}(K)\right)$.

The Axiom$^S$ 1 presented below is a kind of mathematical generalization of Axiom$^P$ 1.

***Axiom$^S$ 1. [Statistical measurement]:*** The probability that a measured value $x(\in X)$ obtained by the measurement $M_A\left(O \equiv (X, \mathcal{F}, F), S_{[*]}(\{\rho_0^m\})\right)$ belongs to a set $\Xi(\in \mathcal{F})$ is given by $\rho_0^m\left(F(\Xi)\right)\left(=_{A^*} \langle \rho_0^m, F(\Xi) \rangle_A\right)$.

Thus, we can propose the statistical measurement theory $(B_2)$, in which Axiom 2 and Interpretation (G) are common.

Let $\hat{O}(X \times Y, \mathcal{F} \boxtimes, H)$ be an observable in a $C^*$- algebra A. Assume that we know that the measured value $(x,y)(\in X \times Y)$ obtained by a statistical measurement $M_A\left(\hat{O}, S_{[*]}\left(\{\rho_0^m\}\right)\right)$ belongs to $\Xi \times Y(\mathcal{F} \boxtimes)$.

Then, there is a reason to infer that the unknown measured value $y(\in Y)$ is distributed under the conditional probability $P_\Xi\left(G(\Gamma)\right)$, where

$$P_\Xi\left(G(\Gamma)\right) = \frac{_A{}^*\left\langle \rho_0^m, H(\Xi \times \Gamma)\right\rangle_A}{_A{}^*\left\langle \rho_0^m, H(\Xi \times Y)\right\rangle_A} \quad (\forall \Gamma \in \mathcal{G})$$

(4)

Thus, by a hint of Fisher's maximum likelihood method, we have the following theorem, which is the most fundamental in this paper.

***Theorem 1. [Fisher's maximum likelihood method in general A]:*** Let $\hat{O}(X \times Y, \mathcal{F} \boxtimes$ be an observable in a $C^*$-algebra A. Let $K(\subseteq \mathfrak{S}^m(A^*))$ be a compact set. Assume that we know that the measured value $(x,y)(\in X \times Y)$ obtained by a measurement $M_A\left(\hat{O}, S_{[*]}(K)\right)$ belongs to $\Xi \times Y(\mathcal{F} \boxtimes)$. Then, there is a reason to infer that the unknown measured value $y(\in Y)$ is distributed under the conditional probability $P_\Xi\left(G(\Gamma)\right)$, where

$$P_\Xi\left(G(\Gamma)\right) = \frac{_A{}^*\left\langle \rho_0^m, H(\Xi \times \Gamma)\right\rangle_A}{_A{}^*\langle \rho_0^m, H(\Xi \times Y)\rangle_A} \quad (\forall \Gamma \in \mathcal{G})$$

(5)

Here, $\rho_0^m \left( \in K \subseteq \mathfrak{S}^m \left( A^* \right) \right)$ is defined by

$$_{A^*}\left\langle \rho_0^m, H \left( \Xi \times Y \right) \right\rangle_A = \max_{\rho^m \in K_{A^*}} \left\langle \rho^m, H \left( \Xi \times Y \right) \right\rangle_A$$

***Remark 2:*** Theorem 1 is new throughout our research [2-8], though, in a particular case that $K \subseteq \mathfrak{S}^p \left( A^* \right)$, Theorem 1 was proposed in [7] where we devoted ourselves to PMT.

## Our Concern in This Paper

Note that $(H_1)$ $M_A \left( O, S_{[\rho_0^p]} \right) = M_A \left( O, S_{[*]} \left( \left\{ \rho_0^p \right\} \right) \right)$ for $\rho_0^p \in \mathfrak{S}^p \left( A^* \right)$, therefore, we see that [PMT] $\subseteq$ [SMT].

However, we have the following problem:

$(H_2)$ Which is fundamental, PMT or SMT?

Recalling the (C), most readers may consider that PMT is more fundamental than SMT. In fact, throughout our research [2-8], we have believed in the fundamentality of PMT. However, in this paper, we assert that Theorem 1 in SMT is the most fundamental as far as inference. In fact, every result in this paper is regarded as one of the corollaries of Theorem 1. And hence, we shall conclude that SMT is proper as the answer to the problem (A). Also, our proposal has a merit such that the philosophy of statistics is naturally induced by the philosophy of measurement theory (cf. [8]).

# FISHER-BAYES METHOD IN CLASSICAL $C(\Omega)$

## Notations

We shall devote ourselves to classical case (i.e. $A = C_0(\Omega)$). From here, $C_0(\Omega)$ (or, commutative unital $C^*$-algebra that includes $C_0(\Omega)$) is, for simplicity, denoted by $C_0(\Omega)$. Thus, we put

$$A^* = C(\Omega)^* = \mathcal{M}(\Omega),$$

$$\mathfrak{S}^m\left(A^*\right) = \mathfrak{S}^m\left(C(\Omega)^*\right) = \mathcal{M}_{+1}^m(\Omega),$$

and

$$\mathfrak{S}^p\left(A^*\right) = \mathfrak{S}^p\left(C(\Omega)^*\right) = \mathcal{M}_{+1}^p(\Omega) \approx \Omega$$

.

And, for any mixed state $v \in M_{+1}^m(\Omega)$ and any observable $O \equiv (X, \mathcal{F}, F)$ in $C(\Omega)$, we put:

$$v\left(F(\Xi)\right) = {}_{C(\Omega)^*}\left\langle v, F(\Xi)\right\rangle_{C(\Omega)} = {}_{\mathcal{M}(\Omega)}\left\langle v, F(\Xi)\right\rangle_{C(\Omega)}$$

$$= \int_\Omega \left[F(\Xi)\right](\omega)v(d\omega).$$

(6)

**Also, put** $v(D) = \int_D v(d\omega)$ $(\forall D \in \beta_\Omega : \text{Boral}\,\sigma - \text{fileld})$: **In order to avoid the confusion between** $v(F(\Xi))$ **in (6) and** $v(D)$, **we do not use** $v(F(\Xi))$. **Also, for any** $\delta_{\omega_0} \in M_{+1}^p(\Omega) \approx \Omega$, **we put:**

$$_{C(\Omega)^*}\left\langle \delta_{\omega_0}, F(\Xi) \right\rangle_{C(\Omega)} =_{\mathcal{M}(\Omega)} \left\langle \delta_{\omega_0}, F(\Xi) \right\rangle_{C(\Omega)}$$

$$= \int_\Omega \left[ F(\Xi) \right](\omega) \delta_{\omega_0}(d\omega) = \left[ F(\Xi) \right](\omega_0).$$

## Bayes Method in Classical $C(\Omega)$

Let $O_1 \equiv (X, \mathcal{F}, F)$ be an observable in a commutative $C^*$-algebra $C(\Omega)$. And let $O_2 \equiv (Y, \mathcal{G}, G)$ be any observable in $C(\Omega)$. Consider the product observable $O_1 \times O_2 \equiv (X \times Y, \mathcal{F} \boxtimes, \mathcal{F}, F \times G)$ in $C(\Omega)$. The existence will be shown in Section 7 (Appendix).

Assume that we know that the measured value $(x, y)$ obtained by a simultaneous measurement $M_{C(\Omega)}\left(O_1 \times O_2, S_{[*]}(\{v_0\})\right)$ belongs to $\Xi \times Y \in \mathcal{F} \boxtimes \mathcal{G}$). Then, by (4), we can infer that

(I) the probability $P_\Xi(G(\Gamma))$ that y belongs to $\Gamma \in$) is given by

$$P_\Xi(G(\Gamma)) = \frac{\int_\Omega \left[ F(\Xi) G(\Gamma) \right](\omega) v_0(d\omega)}{\int_\Omega \left[ F(\Xi) \right](\omega) v_0(d\omega)} \quad (\forall \Gamma \in \mathcal{G})$$

Thus, we can assert that:

**Theorem 2: [Bayes method, cf. [4, 5]]:** When we know that a measured value obtained by a measurement $M_{C(\Omega)}\left(O_1 \equiv (X, \mathcal{F}, F), S_{[*]}(\{v_0\})\right)$ belongs to $\Xi$, there is a reason to infer that the mixed state after the measurement is equal to $v_0^a\left(\in M_{+1}^m(\Omega)\right)$, where

$$v_0^a(D) = \frac{\int_D [F(\Xi)](\omega) v_0(d\omega)}{\int_\Omega [F(\Xi)](\omega) v_0(d\omega)} \quad (\forall D \in \mathcal{B}_\Omega)$$

**Proof:** Note that we can regard that

$P_\Xi \in M_{+1}^m(\Omega)\left(\subseteq C(\Omega)^*\right)$. That is, there exists $v_0^a\left(\varepsilon C(\Omega)^*\right)$ such that

$$P_\Xi(G(\Gamma)) = \int_\Omega [G(\Gamma)](\omega) v_0^a(d\omega) \, (\forall \Gamma \in \mathcal{G})$$

(7)

Then, Axiom[s] 1 says that the probability that a measured value $y(\in Y)$ obtained by the measurement $M_{C(\Omega)}\left(O_2 \equiv (Y, \mathcal{G}, G), S_{[*]}(\{v_0^a\})\right)$ belongs to a set $\Gamma(\in \mathcal{G})$ is given by $\int_\Omega [G(\Gamma)](\omega) v_0^a(d\omega)$, which is equal to $P_\Xi(G(\Gamma))$ in (7). Since $O_2 \equiv (Y, ,G)$ is arbitrary, we obtain Theorem 2.

**Remark 3:** The above (I) is, of course, fundamental. However, in the sense mentioned in the above proof, we admit Theorem 2 as the equivalent statement of the (I). That is, in spite of Interpretation $(G_2)$, we admit the wave function collapse such as (J)

(pretest state)            (posttest state)

$$v_0 \quad \xrightarrow[\text{Theorem 2}]{\text{Bayes}} \quad v_0^a$$

$\left(\in M_{+1}^m(\Omega)\right)$       $\left(\in M_{+1}^m(\Omega)\right)$

Theorem 2 was, for the first time, proposed in [4, 5] without the conscious understanding of Interpretation $(G_2)$. Also, note that (K) in Theorem 2, if $v_0 = \delta_{\omega_0} \left( \in M_{+1}^p (\Omega) \right)$, then it clearly holds that $v_0^a = \delta_{\omega_0}$.

Also, for our opinion concerning the wave function collapse in quantum mechanics, see [7].

## Fisher-Bayes Method in Classical C ($\Omega$)

Combining Theorem 1 (Fisher's method) and Theorem 2 (Bayes' method), we get the following corollary.

**Corollary 1:** [Fisher-Bayes method (i.e., Regression analysis in a narrow sense)]. When we know that a measured value obtained by a measurement $M_{C(\Omega)} \left( O_1 \equiv (X, \mathcal{F}, F), S_{[*]} (K) \right)$ belongs to $\Xi$, there is a reason to infer that the state after the measurement is equal to $v_0^a \left( \in M_{+1}^m (\Omega) \right)$ such that

$$v_0^a (D) = \frac{\int_D \left[ F(\Xi) \right] (\omega) v_0 (d\omega)}{\int_\Omega \left[ F(\Xi) \right] (\omega) v_0 (d\omega)} \quad (\forall D \in \mathcal{B}_\Omega)$$

Where the $v_0 (\in K)$ is defined by

$$\int_\Omega \left[ F(\Xi) \right] (\omega) v_0 (d\omega) = \max_{v \in K} \int_\Omega \left[ F(\Xi) \right] (\omega) v (d\omega)$$

**Remark 4:** As mentioned in the above, note that Corollary 1 is composed of the following two procedure: (L)

$$K \xrightarrow[\text{Theorem 1}]{\text{Fisher}} V_0 \xrightarrow[\text{Theorem 2}]{\text{Bayes}} V_0^a$$
$$\left(\subseteq M_{+1}^m(\Omega)\right) \quad (\in K) \qquad \in K$$

## A Simple Example of Fisher-Bayes Method (Regression Analysis in a Narrow Sense)

In this section, we examine Corollary 1 in a simple example. Readers will find that Corollary 1 can be regarded as regression analysis in a narrow sense.

We have a rectangular water tank filled with water. Assume that the height of water at time t is given by the following function $h(t)$:

$$h(t) = \alpha_0 + \beta_0 t \tag{8}$$

Where $\alpha_0$ and $\beta_0$ are unknown fixed parameters such that $\alpha_0$ is the height of water filling the tank at the beginning and $\beta_0$ is the increasing height of water per unit time. The measured height $h_m(t)$ of water at time t is assumed to be represented by

$$h_m(t) = \alpha_0 + \beta_0 t + e(t) \tag{9}$$

Where $e(t)$ represents a noise (or more precisely, a measurement error) with some suitable conditions. And assume that we obtained the measured data of the heights of water at t=0, 1, and 2 as follows:

$$h_m(0) = 0.5, \quad h_m(1) = 1.6, \quad h_m(2) = 3.3 \tag{10}$$

**A Measurement Theoretical Foundation of Statistics**

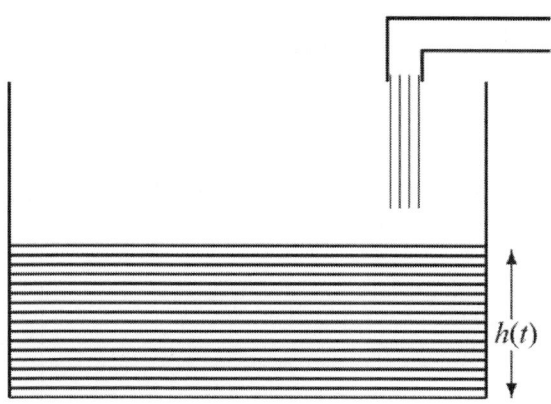

Under this setting, we shall study the following problem:

(M) [Inference]: when measured data (10) is obtained, infer the unknown parameter $(\alpha_0, \beta_0)$ in (9).

In what follows, from the measurement theoretical point of view, we shall answer the problem (M). Let $T = \{0, 1, 2\}$ be a series ordered set such that the parent map $\pi : T \setminus \{0\} \to T$ is defined by $\pi(t) = t - 1$. $(t = 0, 1, 2)$.

Put $\Omega_0 = [0, 2] \times [0, 2]$, $\Omega_1 = [0, 4] \times [0, 2]$, $\Omega_2 = [0, 6] \times [0, 2]$. For each $t = 1, 2$, consider a continuous map $\phi_{\pi(t), t} : \Omega_{\pi(t)} \to \Omega_t$ such that

$$\phi_{0,1}(\alpha, \beta) = (\alpha + \beta, \beta)(\forall \omega_0 = (\alpha, \beta) \in \Omega_0)$$
$$\phi_{1,2}(\alpha, \beta) = (\alpha + \beta, \beta)(\forall \omega_1 = (\alpha, \beta) \in \Omega_1).$$

(11)

Then, we get the deterministic causal operators hus, $\{\Phi_{\pi(t), t} : C(\Omega_t) \to C(\Omega_{\pi(t)})\}_{t \in \{1, 2\}}$ such that

$$\left(\Phi_{0,1}f_1\right)(\omega_0) = f_1\left(\phi_{0,1}(\omega_0)\right)\left(\forall f_1 \in C(\Omega_1), \forall \omega_0 \in \Omega_0\right)$$
$$\left(\Phi_{1,2}f_2\right)(\omega_1) = f_2\left(\phi_{1,2}(\omega_1)\right)\left(\forall f_2 \in C(\Omega_2), \forall \omega_1 \in \Omega_1\right).$$

(12)

Thus, we have the causal relation as follows.

$$C(\Omega_0) \xleftarrow{\quad \Phi_{0,1} \quad} C(\Omega_1) \xleftarrow{\quad \Phi_{1,2} \quad} C(\Omega_2)$$

Put $\phi_{0,2}(\omega_0) = \phi_{1,2}\left(\phi_{0,1}(\omega_0)\right)$, $\Phi_{0,2} = \Phi_{0,1}\cdot\Phi_{1,2}$.

Let $\mathbb{R}$ be the set of real numbers. Fix $\sigma > 0$. For each $t = 0,1,2$, define the normal observable $\left(O_t \equiv \left(\mathbb{R},\mathcal{B}_\mathbb{R},G_\sigma^n\right)\right.$ in $C(\Omega_t)$ such that

$$\left[G_\sigma^n(\Xi)\right](\omega_t) = \frac{1}{\sqrt{2\pi\sigma^2}}\int_\Xi \exp\left(-\frac{(x-\alpha)^2}{2\sigma^2}\right)dx$$
$$\left(\forall \Xi \in \mathcal{B}_\mathbb{R}, \forall \omega_t = (\alpha,\beta) \in \Omega_t = [0, 2t+2]\times[0,2]\right)$$

(13)

Thus, we get the sequential deterministic causal observable

$$[\mathbb{O}_T] = \left[\{O_t\}_{t=0,1,2}, \{\Phi_{\pi(t),t} : C(\Omega_t) \to C(\Omega_{\pi(t)})\}_{t=1,2}\right].$$

Then, the realized causal observable $\hat{O}_0 \equiv \left(\mathbb{R}^3, \mathcal{B}_{\mathbb{R}^3}, \hat{F}_0\right)$ in $C(\Omega_0)$ is, by (3) and (12), obtained as follows:

$$\left[\hat{F}_0\left(\Xi_0 \times \Xi_1 \times \Xi_2\right)\right]\left(\omega_0\right)$$

$$=\left[\left(G_\sigma^n\left(\Xi_0\right)\Phi_{0,1}\left(G_\sigma^n\left(\Xi_1\right)\Phi_{1,2}\left(G_\sigma^n\left(\Xi_2\right)\right)\right)\right)\right]\left(\omega_0\right)$$

$$=\left[G_\sigma^n\left(\Xi_0\right)\right]\left(\omega_0\right)\cdot\left[G_\sigma^n\left(\Xi_1\right)\right]\left(\phi_{0,1}\left(\omega_0\right)\right)$$

$$\cdot\left[G_\sigma^n\left(\Xi_2\right)\right]\left(\phi_{0,2}\left(\omega_0\right)\right)$$

$$\left(\forall\Xi_0,\Xi_1,\Xi_2 \in \mathcal{B}_{\mathbb{R}}, \forall\omega_0 = \left(\alpha,\beta\right) \in \Omega_0\right). \tag{14}$$

Putting $K = M_{+1}^p(\Omega_0)$, we have the measurement $M_{C(\Omega_0)}\left(\hat{O}_0, S_{[*]}\right.$ $\left(M_{+1}^p(\Omega_0)\right)$. Recall the (10), that is, the measured value $(x_0, x_1, x_2)$ obtained by the measurement $M_{C(\Omega_0)}\left(\hat{O}_0, S_{[*]}\left(M_{+1}^p(\Omega_0)\right)\right)$ is equal to

$$\left(0.5, 1.6, 3.3\right)\left(\in \mathbb{R}^3\right) \tag{15}$$

Define the closed interval $\Xi_t \, (t = 0, 2, 3)$ such that

$$\Xi_0 = \left[0.5 - \frac{1}{2N}, 0.5 + \frac{1}{2N}\right],$$

$$\Xi_1 = \left[1.6 - \frac{1}{2N}, 1.6 + \frac{1}{2N}\right],$$

$$\Xi_2 = \left[3.3 - \frac{1}{2N}, 3.3 + \frac{1}{2N}\right],$$

for sufficiently large N. Here, Fisher's method (Theorem 1) says that it suffices to solve the problem.

(N) Find $(\alpha_0, \beta_0)$ such as

$$\max_{(\alpha,\beta)\in\Omega_0} \left[ \hat{F}_0 \left( \Xi_0 \times \Xi_1 \times \Xi_2 \right) \right] (\alpha, \beta)$$

(16)

Putting

$$U\left(x_0, x_1, x_2, \alpha, \beta\right) = \sum_{k=0}^{2} \left(x_k - \left(\alpha + k\beta\right)\right)^2$$

We have the following problem that is equivalent to (N):

(O) Find $(\alpha_0, \beta_0)$ such as

$$\min_{(\alpha,\beta)\in\Omega_0} \exp\left( -\frac{U\left(x_0, x_1, x_2, \alpha, \beta\right)}{2\sigma^2} \right)$$

$$\Leftrightarrow \max_{(\alpha,\beta)\in\Omega_0} U\left(x_0, x_1, x_2, \alpha, \beta\right).$$

Calculating

$$\frac{\partial}{\partial\alpha} U\left(0.5, 1.6, 3.3, \alpha, \beta\right) = 0,$$

$$\frac{\partial}{\partial \beta} U(0.5, 1.6, 3.3, \alpha, \beta) = 0,$$

We get

$$(\alpha, \beta) = (0.4, 1.4)$$

(17)

Thus, we see, by the statement (K), that (P)

$$M_{+1}^{p}(\Omega_0) \xrightarrow[\text{Theorem 1}]{\text{Fisher}} \delta_{(0.4,1.4)} \xrightarrow[\text{Theorem 2}]{\text{Bayes}} \delta_{(0.4,1.4)}$$
$$\left(\subseteq M_{+1}^{m}(\Omega)\right) \qquad (\in K) \qquad (\in K)$$

This (i.e., $(\alpha_0, \beta_0) = (0, 4, 1.4)$) is the answer to the problem (M).

**Problem 1:** Since the above example is quite easy, the validity of Bayes' theorem in (P) may not be clear. If it is so, instead of the problem (M), we should present the following simple problem.

(Q) Infer the water level at time 1.

Some may calculate and conclude as follows:

$$h(1) = \alpha_0 + \beta_0 \times 1 = 0.4 + 1.4 = 1.8$$

(18)

However, this calculation is based on the Schrödinger picture, and thus, the justification of this calculation (18) is not assured. That is because measurement theory (particularly, Interpretation $(G_2)$) says that the Heisenberg picture should be adopted. Therefore, in order to answer the problem (Q), we must prepare Corollary 2 (i.e., regression analysis in a wide sense) in the following section.

*Remark 5:* It should be noted that the following two are equivalent:

($R_1$) [= (M); Inference]: when measured data (10) is obtained, infer the unknown parameter $(\alpha_0, \beta_0)$.

($R_2$) [Control]: Settle the parameter $(\alpha_0, \beta_0)$ such that measured data (10) will be obtained.

That is, we see that

"Inference" = "control".

Hence, from the measurement theoretical point of view, we consider that

"Statistics" = "Dynamical system theory" though these are superficially different in applications.

## CAUSAL FISHER-BAYES METHOD IN CLASSICAL C ($\Omega$)

## Causal Bayes Method in Classical $C(\Omega)$

Let $t_0$ be the root of a tree T. Let

$$\left[\mathbb{O}_T^\times\right] = \left[\left\{O_t^\times \left(\equiv \left(X_t \times Y_t, \mathcal{F}_t \boxtimes \mathcal{G}_t, F_t \times G_t\right)\right)\right\}_{t \in T},\right.$$

$$\left.\left\{\Phi_{t_1, t_2} : C\left(\Omega_{t_2}\right) \to C\left(\Omega_{t_1}\right)\right\}_{(t_1, t_2) \in T_\leq^2}\right]$$

be a sequential causal observable with the realization $\hat{O}^{\times}_{t_0}\left(\times_{t\in T}(X_t, Y_t)\right),$ $\boxtimes_{t\in T}\left(\mathcal{F}, \boxtimes_t\right)\hat{H}_{t_0}$ in. $C(\Omega_{t_0})$ Thus we have the statistical measurement $M_{C(\Omega_{t_0})}\left(\hat{O}, S_{[*]}(\{v_0\})\right)$, where $v_0 \in M^m_{+1}(\Omega_{t_0})$. Assume that we know that the measured value $(x,y)\left(=(x_t)_{t\in T},(x_t)_{t\in T}\right)\in\left(\times_{t\in T}X_t\right)\times\left(\times_{t\in T}Y_t\right)$ obtained by the measurement $M_{C(\Omega_{t_0})}\left(\hat{O}, S_{[*]}(\{v_0\})\right)$ belongs to $\left(\times_{t\in T}\Xi_t\right)\times\left(\times_{t\in T}Y_t\right)\left(\in(\boxtimes_{t\in t}\mathcal{F}_t)\boxtimes(\boxtimes_{t\in t}Y_t)\right)$. Then, by (4), we can infer that (S) the probability $P_{\times_{t\in T}\Xi_t}\left(\left(G_t(\Gamma_t)\right)_{t\in T}\right)$ that y belongs to $\times_{t\in T}\Gamma(\in\boxtimes_{t\in T})$ is given by

$$
P_{\times_{t\in T}\Xi_t}\left(\left(G_t(\Gamma_t)\right)_{t\in T}\right)
$$
$$
=\frac{\int_\Omega\left[\hat{H}_{t_0}\left(\left(\times_{t\in T}\Xi_t\right)\times\left(\times_{t\in T}\Gamma_t\right)\right)\right](\omega)v_0(d\omega)}{\int_\Omega\left[\hat{H}_{t_0}\left(\times_{t\in T}\Xi_t\right)\times\left(\times_{t\in T}Y_t\right)\right](\omega)v_0(d\omega)}
$$
$$
\left(\forall\Gamma_t\in\mathcal{G}_t, t\in T\right).
$$
$$\tag{19}$$

Note that we can regard that $P_{\times_{t\in T}\Xi_t}\in M^m_{+1}\left(\times_{t\in T}\Omega_t\right)\left(\subseteq C\left(\times_{t\in T}\Omega_t\right)^*\right)$. That is, there uniquely exists $v^a_T\in M^m_{+1}\left(\times_{t\in T}\Omega_t\right)$ such that

$$
P_{\times_{t\in T}\Xi_t}\left(\left(G_t(\Gamma_t)\right)_{t\in T}\right)=\int_{\times_{t\in T}\Omega_t}\left[\otimes_{t\in T}G_t(\Gamma_t)\right](\omega)v^a_T(d\omega)
$$
$$\tag{20}$$

for any observable, $G$) in $C(\Omega_t)$ $(t\in T)$. Here, we used the following notation:

$$\left[ \otimes_{t \in T} G_t\left(\Gamma_t\right)\right](\omega) = \underset{t \in T}{\times}\left[ G_t\left(\Gamma_t\right)\right](\omega_t)$$

$$\left(\forall \omega = \left(\omega_t\right)_{t \in T} \in \underset{t \in T}{\times} \Omega_t\right).$$

Define the observable $\hat{O}_{t_0} \equiv \left(\times_{t \in T} X_t, \boxtimes_{t \in T} F_t, \hat{F}_{t_0}\right)$ such that

$$\hat{F}_{t_0}\left(\underset{t \in T}{\times} \Xi_t\right) = \hat{H}_{t_0}\left(\left(\underset{t \in T}{\times} \Xi_t\right) \times \left(\underset{t \in T}{\times} Y_t\right)\right)$$

Then, we can define the Bayes operator $\left[ B_{\hat{O}_{t_0}}\left(\times_{t \in T} \Xi_t\right)\right] : M_{+1}^m\left(\Omega_{t_0}\right) \to M_{+1}^m\left(\times_{t \in T} \Omega_t\right)$ by (20).

Thus, as the generalization of Theorem 2, we have:

***Theorem 3: [Causal Bayes' theorem in classical measurements]:*** Let $t_0$ be the root of a tree T. Let

$$\left[\mathbb{O}_T\right] = \left[\left\{O_t\left(\equiv\left(X_t, \mathcal{F}_t, F_t\right)\right)\right\}_{t \in T},\right.$$

$$\left.\left\{\Phi_{t_1, t_2} : C\left(\Omega_{t_2}\right) \to C\left(\Omega_{t_1}\right)\right\}_{(t_1, t_2) \in T_{\le}^2}\right]$$

be a sequential causal observable with the realization $\hat{O}_{t_0} \equiv \left(\times_{t \in T} X_t, \boxtimes_{t \in T} F_t, \hat{F}_{t_0}\right)$. Thus we have the statistical measurement $M_{C(\Omega_{t_0})}\left(\hat{O}_{t_0}, S_{[*]}\left(\{\nu_0\}\right)\right)$, where $\nu_0 \in M_{+1}^m\left(\Omega_{t_0}\right)$. Assume that we know that a measured value obtained by the statistical measurement $M_{C(\Omega_{t_0})}\left(\hat{O}_{t_0}, S_{[*]}\left(\{\nu_0\}\right)\right)$ belongs to $\times_{t \in T} \Xi_t$. Then, there is a reason to infer that the mixed state $v_T^a \left( \in M_{+1}^m \left( \times_{t \in T} \Omega_t \right) \right)$ after the statistical measurement

$M_{C(\Omega_{t_0})} \left( \hat{O}_{t_0}, S_{[*]} \left( \{ v_0 \} \right) \right)$ is given by

$$\left[ B_{\hat{O}_{t_0}} \left( \times_{t \in T} \Xi_t \right) \right] \left( v_0 \right) \left( \in M_{+1}^m \left( \times_{t \in T} \Omega_t \right) \right).$$

**Proof:** The proof is similar to the proof of Theorem 2. Thus, we omit it.

**Remark 6:** In Theorem 3, we see that (T)

(pretest state)         (posttest state)

$$\begin{array}{ccc}
& \text{Bayes} & v_T^a \\
v_0 & \xrightarrow{\hspace{2cm}} & \\
\left( \in M_{+1}^m \left( \Omega_{t_0} \right) \right) & \text{Theorem 2} & \left( \in M_{+1}^m \left( \times_{t \in T} \Omega_t \right) \right)
\end{array}$$

Which is the generalization of the (J).

The following example promotes the understanding of Theorem 3.

**Example 1. [The simple case such that** $T = \{0,1,2\}$ **]:** Consider a particular case such that $T = \{0,1,2\}$ is series ordered set, i.e., $\left( \forall t \in T \setminus \{0\} \right) \pi(t) = t - 1$. And consider a causal relation

$\left\{ C(\Omega_t) \xrightarrow{\Omega_{\pi(t),t}} C\left( \Omega_{\pi(t)} \right) \right\}_{t \in T \setminus \{0\}}$ , that is,

$$C(\Omega_0) \xleftarrow{\Phi_{0,1}} C(\Omega_1) \xleftarrow{\Phi_{1,2}} C(\Omega_2) .$$

Further consider sequential causal observable

$$\left[\mathbb{O}_T\right] = \left[\left\{\mathsf{O}_t\right\}_{t \in T}, \left\{\Phi_{t,\pi(t)} : C(\Omega_t) \to C\left(\Omega_{\pi(t)}\right)\right\}_{t \in T \setminus \{0\}}\right].$$

Let $\hat{\mathsf{O}}_0 \equiv \left(\times_{t \in T} X_t, \times_{t \in T} F_t, \hat{F}_0\right)$ be its realization. Note, by the Formula (3), that,

$$\hat{F}_0\left(\Xi_0 \times \Xi_1 \times \Xi_2\right)$$
$$= \Phi_{0,1}\left(F_0\left(\Xi_0\right)\left(\Phi_{1,2}F_1\left(\Xi_1\right)\left(\Phi_{1,2}\left(F_2\left(\Xi_2\right)\right)\right)\right)\right)$$
$$\left(\Xi_t \in \mathcal{F}_t\left(\forall t \in T\right)\right).$$

Putting $K = \{v_0\}$, we have the measurement

$$M_{C(\Omega_0)}\left(\hat{\mathsf{O}}_0 \equiv \left(\times_{t \in T} X_t, \boxtimes_{t \in T} F_t, \hat{F}_0\right), S_{[*]}\left(\{v_0\}\right)\right)$$

$$(21)$$

Let $V_T^a\left(\in M_{+1}^m\left(\Omega_0 + \Omega_1 \times \Omega_2\right)\right)$ be the posttest state in (T), that is,

$V_T^a\left[B_{\hat{\mathsf{O}}_0}\left(\times_{t \in T} \Xi_t\right)(v_0)\right]$. Define $V_{\{1\}}^a\left(\in M_{+1}^m\left(\Omega_1\right)\right)$ such that

$$v_{\{1\}}^a\left(D_1\right) = v_T^a\left(\Omega_0 \times D_1 \times \Omega_2\right) \quad \left(\forall D_1 \in \mathcal{B}_{\Omega_1}\right).$$

Then, we see that

$$v_{\{1\}}^a = \frac{\left(F_1\left(\Xi_1\right)\left(\Phi_{1,2}F_2\left(\Xi_2\right)\right)\right)\left(\Phi_{0,1}^*\left(F_0\left(\Xi_0\right)v_0\right)\right)}{\left\langle v_0, F_0\left(\Xi_0\right)\Phi_{0,1}\left(F_1\left(\Xi_1\right)\Phi_{1,2}\left(F_2\left(\Xi_2\right)\right)\right)\right\rangle}$$

That is because we see that, for any observable $(Y_1, G_1)$ in $C(\Omega_1)$,

$$= \frac{\left\langle v_0, F_0(\Xi_0)\Phi_{0,1}\left(F_1(\Xi_1)G_1(\Gamma_1)\Phi_{1,2}(F_2(\Xi_2))\right)\right\rangle}{\left\langle v_0, F_0(\Xi_0)\Phi_{0,1}\left(F_1(\Xi_1)G_1(Y_1)\Phi_{1,2}(F_2(\Xi_2))\right)\right\rangle}$$

$$= \frac{\left\langle \left(F_1(\Xi_1)\left(\Phi_{1,2}F_2(\Xi_2)\right)\right)\left(\Phi_{0,1}^*\left(F_0(\Xi_0)v_0\right)\right), G_1(\Gamma_1)\right\rangle}{\left\langle v_0, F_0(\Xi_0)\Phi_{0,1}\left(F_1(\Xi_1)\Phi_{1,2}(F_2(\Xi_2))\right)\right\rangle} \tag{22}$$

$(\forall \Gamma_1 \in \mathcal{G}_1)$.

***Example 2. [Continued from the above example]:*** For each $t = 1, 2$, assume that $\Phi_{\pi(t),t} : C(\Omega_1) \to C(\Omega_{\pi(t)})$ is deterministic, that is, there exists a continuous map $\phi_{\pi(t),t} : \Omega_{\pi(t)} \to \Omega_t$ satisfying (12). And, putting $K = \{\delta_{\omega_0}\}$, consider the measurement

$$M_{C(\Omega_0)}\left(\hat{O}_0 \equiv \left(\times_{t\in T} X_t, \boxtimes_{t\in T} F_t, \hat{F}_0\right), S_{[*]}\left(\{\delta_{\omega_0}\}\right)\right).$$

Then, we see, by (22), that, for any $g_1$ in $C(\Omega_1)$,

$$\left\langle v_{\{1\}}^a, g_1\right\rangle = \frac{\left\langle \delta_{\omega_0}, F_0(\Xi_0)\Phi_{0,1}\left(F_1(\Xi_1)g_1\Phi_{1,2}(F_2(\Xi_2))\right)\right\rangle}{\left\langle \delta_{\omega_0}, F_0(\Xi_0)\Phi_{0,1}\left(F_1(\Xi_1)\Phi_{1,2}(F_2(\Xi_2))\right)\right\rangle}$$

$$= \frac{\left[F_0(\Xi_0)\right](\omega_0)\left[F_1(\Xi_1)g_1\Phi_{1,2}(F_2(\Xi_2))\right](\phi_{0,1}(\omega_0))}{\left[F_0(\Xi_0)\right](\omega_0)\left[F_1(\Xi_1)\Phi_{1,2}(F_2(\Xi_2))\right](\phi_{0,1}(\omega_0))}$$

$$= g_1\left(\phi_{0,1}(\omega_0)\right) = \left\langle \delta_{\phi_{0,1}(\omega_0)}, g_1\right\rangle.$$

Thus, we see that

$$v_{\{1\}}^a = \delta_{\phi_{0,1}(\omega_0)}$$

$$(23)$$

Further we easily see that

$$v_T^a = \left[ B_{\hat{O}_0} \left( \times_{t \in T} \Xi_t \right) \right] \left( \delta_{\omega_0} \right)$$

$$= \delta_{\left( \omega_0, \phi_{0,1}(\omega_0), \phi_{0,2}(\omega_0) \right)} \left( \in \mathcal{M}_{+1}^p \left( \Omega_0 \times \Omega_1 \times \Omega_2 \right) \right).$$

## Causal Fisher-Bayes Method in Classical C ($\Omega$)

Now we can present Corollary 2 (i.e., regression analysis in a wide sense) as follows (U)

$$\left[ \text{Corollary 2} \right] = \underset{\text{(Fisher's method)}}{\left[ \text{Theorem 1} \right]} + \underset{\text{(Bayes' method)}}{\left[ \text{Theorem 3} \right]}$$

*Corollary 2. [Causal Fisher-Bayes method (i.e., Regression analysis in a wide sense)]:* Let $t_0$ be the root of a tree T. Let

$$\left[ \mathbb{O}_T \right] = \left[ \left\{ O_t \left( \equiv \left( X_t, \mathcal{F}_t, F_t \right) \right) \right\}_{t \in T}, \right.$$

$$\left. \left\{ \Phi_{t_1, t_2} : C \left( \Omega_{t_2} \right) \rightarrow C \left( \Omega_{t_1} \right) \right\}_{(t_1, t_2) \in T_{\leq}^2} \right]$$

be a sequential causal observable with the realization $\hat{O}_{t_0} \equiv (\times_{t \in T} X_t, \boxtimes_{t \in T} F_t, \hat{F}_{t_0})$. Assume the statistical measurement $M_{C(\Omega_{t_0})} (\hat{O}_{t_0}, S_{[*]}(K))$. And assume that we know that a measured value obtained by the measurement $M_{C(\Omega_{t_0})} (\hat{O}_{t_0}, S_{[*]}(K))$ belongs to $\times_{t \in T} \Xi_t$. Then, there is a reason to infer that the mixed state $V_T^a (\in M_{+1}^m (\times_{t \in T} \Omega_t))$ after the measurement

$M_{C(\Omega_{t_0})}\left(\hat{O}_{t_0}, S_{[*]}(K)\right)$ is given by $\left[B_{\hat{O}_{t_0}}\left(\times_{t\in T}\Xi_t\right)\right](v_0)$. Here, the $v_0\ (\in K)$ is defined by

$$\int_\Omega\left[\hat{F}_{t_0}\left(\times_{t\in T}\Xi_t\right)\right](\omega)v(d\omega)$$

$$= \max_{v\in K}\int_\Omega\left[\hat{F}_{t_0}\left(\times_{t\in T}\Xi_t\right)\right](\omega)v(d\omega)$$

(24)

**Remark 7:** Note that Fisher maximum likelihood method and Bayes' theorem are hidden in Corollary 2. That is, Corollary 2 includes the following procedure: (V)

$$\underset{\left(\in\mathcal{M}_{+1}^m\left(\Omega_{t_0}\right)\right)}{K} \xrightarrow[\text{Theorem 1}]{\text{Fisher}} \underset{(\in K)}{V_0} \xrightarrow[\text{Theorem 3}]{\text{Bayes}} \underset{\left(\in\mathcal{M}_{+1}^m\left(\times_{t\in T}\Omega_t\right)\right)}{V_T^a}$$

Which is the generalization of the (L).

**Answer 1. [Answer to Problem 1 (Q)]:** Now we can answer Problem 1 (Q) as follows. The (17) says that

$$V_0 = \delta_{(\alpha_0,\beta_0)} = \delta_{(0.4,1.4)}.$$ Thus, using (23), we see that

$$V_{\{1\}}^a = \delta_{(\alpha_0,\beta_0)} = \delta_{1.8}.$$ Also, note that (17) and (23) are consequences of Corollary 2. Hence, the calculation (18) is justified by Corollary 2.

**Remark 8:** As mentioned in Section 1, in our research [2-8], we have been concerned with the problem (A). Particularly, in [6], we discussed Corollary 2 in the commutative W*-algebra $L^\infty(\Omega)$. However, this was somewhat shallow, since "max" is not proper in $L^\infty(\Omega)$ but $C(\Omega)$. Now we believe that fundamental statements concerning statistics should be always asserted in the framework of $C(\Omega)$. Also, note that Corollary 2 is the natural generalization of Theorem 6.3 in [5].

## CONCLUSIONS

In this paper, we devote ourselves to the problem (A) in the light of the quantum mechanical word view (cf. [7, 8]). And, we show that regression analysis, which is the most fundamental in statistics, is formulated as Corollary 2 in SMT (i.e., statistical measurement theory). We believe that Corollary 2 is the finest formulation of regression analysis, since no clear formulation can be presented without the answer to the problem (A). Also, note that Corollary 2 (or, the (U)) implies that even the conventional classification of (Fisher's) statistics and Bayesian statistics should be reconsidered.

We expect that there is a great possibility that our proposal (i.e., statistics is based on statistical measurement theory) will be generally accepted. We of course know that the conventional statistics methodology can be good applied in many fields. Hence, we hope that our methodology in the light of the quantum mechanical word view should be examined from various points of view.

## APPENDIX

As mentioned in Section 3.1, we have to prove the following theorem.

***Theorem 4. [Existence theorem of product observable]:*** Let $O_1 \equiv (X, \mathcal{F}, F)$ and $O_2 \equiv (Y, \mathcal{G}, G)$ be observables in a C*-algebra $A = C(\Omega)$. Then, there exists the product observable $O_1 \times O_2 \equiv (X \times Y, \mathcal{F} \boxtimes \mathcal{G}, F \times G)$ in $C(\Omega)$.

**Proof:** Let $\bar{\mathcal{F}}$ [resp. $\bar{\mathcal{G}}$; $\bar{\mathcal{F}} \boxtimes \bar{\mathcal{G}}$] be the smallest $\sigma$-field including $\mathcal{F}$ [resp. $\mathcal{G}$; $\mathcal{F} \boxtimes \mathcal{G}$]. That is, for each k=1, 2... consider $(\in \mathcal{F} \boxtimes \mathcal{G}) \Xi_k \times \Gamma_k$ such that

$$\left( \Xi_i \times \Gamma_i \right) \cap \left( \Xi_j \times \Gamma_j \right) = \emptyset \, (i \neq j)$$

and

$$\bigcup_{k=1}^{\infty} \left( \Xi_k \times \Gamma_k \right) \in \mathcal{F} \boxtimes \mathcal{G}.$$

Note, by the Hopf extension theorem (cf. Remark 1), that it suffices to show that, for any $v \in M_{+1}^m(\Omega)$, it holds:

$$\int_{\Omega} \left[ (F \times G) \left( \bigcup_{k=1}^{\infty} (\Xi_k \times \Gamma_k) \right) \right] (\omega) v(d\omega)$$

$$= \lim_{K \to \infty} \int_{\Omega} \sum_{k=1}^{K} \left[ F(\Xi_k) \right] (\omega) \cdot \left[ G(\Gamma_k) \right] (\omega) v(d\omega)$$

Which is equivalent to the following equality. That is, for any $\omega \in \Omega$, it holds:

$$\left[ (F \times G) \left( \bigcup_{k=1}^{\infty} (\Xi_k \times \Gamma_k) \right) \right] (\omega)$$

$$= \lim_{K \to \infty} \sum_{k=1}^{K} \left[ F(\Xi_k) \right] (\omega) \cdot \left[ G(\Gamma_k) \right] (\omega). \tag{25}$$

However, it is easily seen since $(X, \bar{\mathcal{F}}, [F(\cdot)](\omega))$ and $(Y, \bar{\mathcal{G}}, [G(\cdot)](\omega))$ can be regarded as probability spaces. And therefore, we have the product probability space $(X \times Y, \bar{\mathcal{F}} \boxtimes \bar{\mathcal{G}}, [(F \times G)(\cdot)](\omega))$. This imlies that the equality (25) holds. This completes the proof.

***Remark 9:*** The above proof is applicable to the realization of a sequential causal observable $[\mathbb{O}_T]$ in the case of an infinite T under a similar condition such that the Kolmogorov extension theorem holds (cf. [1]). Also, in quantum case (i.e., $A=B_c(H)$), it is well known that the weak convergence (1) in $B_c(H)$ can be identified with the weak convergence in B(H), therefore, we see, by a usual way (cf. [10,11]), that Theorem 4 holds under the commutativity condition (2).

## REFERENCES

1. Kolmogorov, "Foundations of the Theory of Probability (Translation)," 2nd Edition, Chelsea Pub Co., New York, 1960,

2. S. Ishikawa, "Fuzzy Inferences by Algebraic Method," Fuzzy Sets and Systems, Vol. 87, No. 2, 1997, pp. 181- 200. doi:10.1016/S0165-0114(96)00035-8

3. S. Ishikawa, "A Quantum Mechanical Approach to Fuzzy Theory," Fuzzy Sets and Systems, Vol. 90, No. 3, 1997, pp. 277-306. doi:10.1016/S0165-0114(96)00114-5

4. S. Ishikawa, "Statistics in Measurements," Fuzzy Sets and Systems, Vol. 116, No. 2, 2000, pp. 141-154. doi:10.1016/S0165-0114(98)00280-2

5. S. Ishikawa, "Mathematical Foundations of Measurement Theory," Keio University Press Inc., Tokyo, 2006, pp. 1-335. http://www.keio-up.co.jp/kup/mfomt/

6. S. Ishikawa, "Fisher's Method, Bayes' Method and Kalman Filter in Measurement Theory," Far East Journal of Theoretical Statistics, Vol. 29, No. 1, 2009, pp. 9-23.

7. S. Ishikawa, "A New Interpretation of Quantum Mechanics," Journal of Quantum Information Science, Vol. 1, No. 2, 2011, pp. 35-42. doi:10.4236/jqis.2011.12005

8. S. Ishikawa, "Quantum Mechanics and the Philosophy of Language: Reconsideration of Traditional Philosophies," Journal of Quantum Information Science, Vol. 2, No. 1, 2012, pp. 2-9.

9. G. J. Murphy, "C*-Algebras and Operator Theory," Academic Press, London, 1990.

10. J. von Neumann, "Mathematical Foundations of Quantum Mechanics," Springer Verlag, Berlin, 1932.

11. K. Yosida, "Functional Analysis," 6th Edition, SpringerVerlag, Berlin, 1980.

12. E. B. Davies, "Quantum Theory of Open Systems," Academic Press, London, 1976.

## CITATION

S. Ishikawa, "A Measurement Theoretical Foundation of Statistics," Applied Mathematics, Vol. 3 No. 3, 2012, pp. 283-292. Doi: 10.4236/am.2012.33044.

# Boolean Automorphisms of a Hypercube Coincide with the Linear Isometries

## *Eberto R. Morgado[1] and Marco V. José[2]*

[1]Facultad de Matemática, Física y Computación, Universidad Central "Marta Abreu" de Las Villas, Santa Clara, Cuba

[2]Theoretical Biology Group, Instituto de Investigaciones Biomédicas, Universidad Nacional Autónoma de México, México D.F., México

## ABSTRACT

Boolean homomorphisms of a hypercube, which correspond to the morphisms in the category of finite Boolean algebras, coincide with the linear isometries of the category of finite binary metric vector spaces.

## INTRODUCTION

An automorphism is an isomorphism from a mathematical object to itself. It is, in some sense, symmetry of the object, and a way of mapping the object to itself while preserving all of its structure. The set of all automorphisms of an object forms a group, called the automorphism group. It is, loosely speaking, the symmetry group of the object.

As is well known, a Boolean lattice is a partially ordered set with some special properties of its partial order relation, and it can also be envisaged as an algebraic system with two algebraic binary operations. This algebraic system is the so-called Boolean algebra, associated to the Boolean lattice. The Boolean algebra can also be provided with a ring structure, the so-called Boolean ring, associated to the Boolean lattice. It can even be regarded as a binary vector space, that is, a vector space over the binary field $\mathbb{Z}_2 = \{0,1\}$ of two elements [1]. These four categories are functorial related by isofunctors that carry over the morphisms of one category over morphisms of the other [2]. The Boolean lattice is also a metric space with the so-called Hamming distance, where the morphisms are the so-called isometries.

The aim of the present work is to show that the Boolean homomorphisms, that is, the morphisms in the category of the finite Boolean algebras, are the same to the linear isometries in the category of finite binary vector spaces when the Hamming distance is used.

## SOME PREVIOUS DEFINITIONS AND CONCEPTS

### Definition

Given a Boolean algebra $(B,\vee,\wedge)$ a function $f: B \to B$ such that $f(x \vee y) = f(x) \vee f(y)$, $f(x \wedge y) = f(x) \wedge f(y)$, for all x, y of B, and $f(0) = 0$, $f(1) = 1$, where 0 and 1 denote the neutral elements of $\vee, \wedge$, respectively, is called a Boolean endomorphism. If f is bijective, it is called a Boolean automorphism.

It is immediate that, for the Boolean addition +, defined as $x + y = (x \vee y) \wedge (x' \vee y')$, which provides to B a structure of Abelian group, f is a group endomorphism, or a group authomorphism if it is bijective.

It is well known that the triple $(B,+,\bullet)$ where $\bullet$ denotes the obviously defined external operation of the binary field $\mathbb{Z}_2 = \{0,1\}$ over B, is a $\mathbb{Z}_2$-vector space.

It is not difficult to prove that every group endomorphism of the Abelian group $(B,+)$ is also a linear endomorphism of the vector space $(B,+,\bullet)$

The triplet $(B,+,\wedge)$ is a unitary commutative ring, such that every element x is idempotent, that is, $x \wedge x = x$.

It is easy to notice that every Boolean endomorphism is also a unitary ring endomorphism, and conversely, every unitary ring endomorphism is a Boolean endomorphism.

It is also well known that, the binary relation $\leq$ defined as $x \leq y \Leftrightarrow x \wedge y = x$, is a partial order relation with minimum and maximum 0 and 1, respectively.

The ordered pair $(B, \leq)$ defines a lattice, such that for every binary subset $\{x, y\}$ the elements $x \vee y$ and $x \wedge y$ are, respectively, the least upper bound (supremum) and the greatest lower bound (infimum) of the set $\{x, y\}$

A function $f : B \to B$ is a Boolean endomorphism if, and only if, it is isotonic with respect to the partial order relation $\leq$, that is, if $x \leq y \Rightarrow f(x) \leq f(y)$ for all x, y of B..

If the vector space $(B, +, \bullet)$ has a finite basis of n elements, we will say that the Boolean algebra $(B, \vee, \wedge)$ is finite-dimensional, being the number n its dimension.

It can be proved that every n-dimensional Boolean algebra is isomorphic to the Boolean algebra $((\mathbb{Z}_2)^n, \vee, \wedge)$, where the operations are bitwise induced by the logic operations of disjunction and conjunction, according to the following Table1

The elements 0 and 1 represent, respectively, falsity or veracity of a proposition. The Boolean algebra $((\mathbb{Z}_2)^n, \vee, \wedge)$, is generally called the n-dimensional hypercube. It is due to the fact that in the case $n = 3$ the triplets of zeros and ones are the algebraic representations of the vertexes of a cube, inserted, as a subset, in the 3-dimensional $\mathbb{R}$-vector space $\mathbb{R}$, being $\mathbb{R}$ the field of real numbers.

## The Inner Product in the Hypercube $\left((\mathbb{Z}_2)^n, \vee, \wedge\right)$,

### Definition

For two n-tuples $u = (x_1, x_2, \cdots, x_n)$ and $v = (y_1, y_2, \cdots, y_n)$ of $(\mathbb{Z}_2)^n$ we call scalar product or inner product of u with v the number $\langle u, v \rangle = x_1 y_1 + x_2 y_2 + \cdots x_n y_n$, where the addition and the multiplication are the ordinary operations in the ring $\mathbb{Z}$ of integers.

This inner product is the restriction to the set $(\mathbb{Z}_2)^n$ of the ordinary inner product of the Euclidean n-dimensional $\mathbb{R}$-vector space $\mathbb{R}^n$.

**Table 1**: Logic operations in Boolean algebra

| $\vee$ | 0 | 1 | | $\wedge$ | 0 | 1 |
|---|---|---|---|---|---|---|
| 0 | 0 | 1 | | 0 | 0 | 0 |
| 1 | 1 | 1 | | 1 | 0 | 1 |

If the column matrices $X = \begin{pmatrix} x_1 \\ x_2 \\ \vdots \\ x_n \end{pmatrix}$ and $Y = \begin{pmatrix} y_1 \\ y_2 \\ \vdots \\ y_n \end{pmatrix}$ are the matrix representation of the n-tuples u and v, respectively, the inner product $\langle u, v \rangle$

can be expressed as the matrix product $^tXY$, where $^tX$ denotes the transpose matrix of X, that is, the row matrix $(x_1, x_2, \cdots, x_n)$.

## Definition

For a vector $u = (x_1, x_2, \cdots, x_n)$. we call the norm, absolute value, or weight of u, the inner product $\langle u, v \rangle = x_1x_1 + x_2x_2 + \ldots + x_nx_n$ of u with itself, denoted as $|u|$. obviously, the norm $|u|$ is equal to the number of times the number 1 is a component of u.

It is not difficult to notice that the inner product $\langle u, v \rangle$ of the vector u and v, is equal to the norm of the vector product $u \wedge v$ in the Boolean ring $((\mathbb{Z}_2)^n, +, \wedge)$,

A vector u of norm $|u| = 1$ is called unitary vector. The only unitary vectors are

$$e_1 = (1, 0, \cdots, 0), e_2 = (0, 1, \cdots, ), \cdots e_i = (0, 0, \cdots, 1, \cdots, 0), \cdots, e_n = (0, 0, \cdots, n)$$

and they conform the so-called canonical basis $(e_1, e_2, \cdots, e_n)$ of the binary vector space $((\mathbb{Z}_2)^n, +, \bullet)$.

## THE CONCEPT OF ORTHOGONALITY

*wang#_2title:spDefinition*

We say that two vectors u and v are orthogonal or perpendicular if the inner product $\langle u, v \rangle$ is equal to 0.

# THE HAMMING DISTANCE IN THE HYPERCUBE

## wang#_2title:spDefinition

For two vectors u and v, we define the Hamming distance between them, as the norm $|u+v|$ of their Boolean addition. Obviously, it is equal to the number of places where the components of both vectors are different.

## LINEAR ISOMETRIES OF THE HYPERCUBE

### Definition

A function $f:(\mathbb{Z}_2)^n \rightarrow (\mathbb{Z}_2)^n$ is called an isometry if it preserves the distance between points, that is, if

$|u+v| = |f(u)+f(v)|$ for all u,v of the set.

If the isometry f is also a linear transformation, then, the matrix A of f, with respect to the canonical basis $(e_1, e_2, \cdots e_n)$ is an orthogonal matrix, that is, such that $^t AA = I_n$, the identity n×n matrix.

It is clear that a linear isometry also preserves the absolute value of any vector and the inner product of any two vectors.

The hypercube $((\mathbb{Z}_2)^n, +, \bullet)$ can also be envisaged as a graph, where the vertexes or nodes are the n-tuplesand the edges are the binary subsets $\{u,v\}$ such that the distance $|u+v|$ is equal to 1. Two vectors u and v of an edge $\{u,v\}$ that is, such that $|u+v|=1$, are called adjacent points of the hypercube.

It can be proved that the Hamming distance between two points u and v is equal to the minimal length of a path between them, that is, the minimal number of edges for going from one to the other.

## MAIN RESULTS

### Lemma

In the hypercube $(\mathbb{Z}_2)^n, n \geq 2$, the only set of n non-null vectors, which are pairwise orthogonal, is the set $\{e_1, e_2, \cdots, e_n\}$ of the unitary canonical vectors.

### Proof: (Induction over n)

For n = 2 the assertion is trivially true.

Let us suppose that it is true for every $t < n,$, being $n > 2$.

Let $\{a_1, a_2, \cdots a_n\}$ be a set of non-null and pairwise orthogonal vectors in the hypercube $(\mathbb{Z}_2)^n$. To prove that they are all unitary vectors let us suppose that one of them, say $a_1$ is not unitary, that is of norm $|a_1| = k > 1$. Then, the vectors $a_2, \cdots, a_n$ belong to the $(n-k)$-dimensional vector subspace, which is the supplementary orthogonal vector subspace of the line $\{0, a_1\}$. As the vectors $a_2 \cdots, a_n$. are linearly independent, then $n-1 = n-k,$. then $k = 1$ in contradiction with the assumption $k > 1$. Hence, all the vectors of the set are unitary, as we wanted to show.

Now, we are in conditions to carry out the proof of the following.

## Theorem

A function f is a Boolean automorphism if, and only if, it is a linear isometry.

Proof: If f is a Boolean automorphism it means that $f(u \vee v) = f(u) \vee f(v)$, $f(u \wedge v) = f(u) \wedge f(v)$, for all u, v of $(\mathbb{Z}_2)^n$ and $f(0) = 0, f(1) = 1$.

Then, $f(u + v) = f(u) + f(v)$ for the Boolean addition $+$ such that, $x + y = (x \vee y) \wedge (x' \vee y')$. Hence, f is a linear transformation of the vector space.

For canonical vectors $e_i, e_j$ we have that $\langle e_i, e_j \rangle$ if $i \neq j$, and $\langle e_i, e_j \rangle = e_i$ if $i = j$, which means that they are unitary and pairwise orthogonal. From this, we have that $f(e_i) \wedge f(e_j) = f(e_i \wedge e_j) = 0$ if $i \neq j$. and $f(e_i) \wedge f(e_j) = f(e_i)$ if $i = j$. Then, the vectors $f(e_1), f(e_2), \cdots, f(e_n)$ are pairwise orthogonal and, from the lemma, they are unitary vectors. Hence, the linear function f is a permutation of the canonical basis $(e_1, e_2, \cdots, e_n)$. Then, the matrix A of f, with respect to this basis, is orthogonal, that is, such that $^t AA = I_n$. Then, f is a linear isometry of the space, as we wanted to prove.

Conversely, if f is a Boolean isometry, $f(e_i) \wedge f(e_j) = f(e_i \wedge e_j)$ for all i and j, then for

$$u = (x_1, x_2, \cdots, x_n) \text{ and } v = (y_1, y_2, \cdots, y_n), \quad f(u) = \sum_{i=1}^{n} x_i f(e_i),$$

$$f(v) = \sum_{j=1}^{n} y_j f(e_j).$$

Then,

On the other hand, it is known that $u \vee v = u + v + (u \wedge v)$ for all u and v. Then,

$$f(u \vee v) = f(u) + f(v) + f(u) \wedge f(v) = f(u) \vee f(v).$$

Then, we have proved that $f(u \vee v) = f(u) \vee f(v)$, $f(u \wedge v) = f(u) \wedge f(v)$ for all u, v elements of the space.

As f is linear we have $f(0) = 0$ and as for every u, $u \vee u' = 1$ we obtain that $f(1) = f(u) \vee f(u) = 1$.

Then, we have proved that f is a Boolean homomorphism.

## CONCLUDING REMARKS

In this work we have demonstrated that Boolean automorphisms of a hypercube are the same to the linear isometries of finite binary metric spaces taking as a metric the Hamming distance. This fundamental result becomes of much interest when characterizing the symmetry groups of polytopes [3]. The use of the theory of categories imparts novel insights for understanding and generalizing the symmetries of any object. This result is of interest in many areas of research. For example, the representation of the Universal Genetic Code as a 6-dimensional hypercube [4] [5] has permitted to study its evolution by a series of successive symmetry breakings [6].

## ACKNOWLEDGEMENTS

MVJ was financially supported by PAPIIT-IN107112, UNAM, México.

# REFERENCES

1.  Dubreil, P. and Jacotin, M.L. (1961) Lecciones de Algebra Moderna. Editorial Dunod, Francia.

2.  Mitchell, B. (1965) Theory of Categories. Academic Press, New York.

3.  Coxeter, H.S.M. (1973) Regular Polytopes. 3rd Edition. Dover Publication Inc., New York.

4.  José, M.V., Morgado, E.R. and Govezensky, T. (2007) An Extended RNA Code and Its Relationship to the Standard Genetic Code: An Algebraic and Geometrical Approach. Bulletin of Mathematical Biology, 69, 215-243. http://dx.doi.org/10.1007/s11538-006-9119-3

5.  José, M.V., Morgado, E.R., Sánchez, R. and Govesenky, T. (2012) The 24 Possible Algebraic Representations of the Standard Genetic Code in Six or in Three Dimensions. Advanced Studies in Biology, 4, 119-152.

6.  José, M.V., Govesenky, T., García, J.A. and Bobadilla, J.R. (2009) On the Evolution of the Standard Genetic Code: Vestiges of Critical Scale Invariance from the RNA World to Current Prokaryote Genomes. PLoS ONE, 4, e4340.http://dx.doi.org/10.1371/journal.pone.0004340

## CITATION

Morgado, E. and José, M. (2014) Boolean Automorphisms of a Hypercube Coincide with the Linear Isometries.Advances in Pure Mathematics, **4**, 368-372. doi: 10.4236/apm.2014.48047.

# Ex Post Efficient Set Mathematics

## *Christopher Adcock*
Sheffield University Management School, University of Sheffield, Sheffield, UK

**6**

## ABSTRACT

This paper considers efficient set mathematics for the case where the covariance matrix of asset returns is assumed known but ex ante the vector of expected returns is replaced by an estimated or forecast value. It is shown that the ex post mean and variance differ from the standard results. Consequently the maximum Sharpe ratio portfolio also differs from the standard result. However, even with uncertainty about the vector of expected returns, subject to the assumptions made about the joint distribution of actual returns and estimated mean returns, ex post Sharpe ratio maximisers hold the ex post market portfolio. The properties of the zero beta portfolio are similar to the standard results leading to a capital market line. The ex post Capital Asset Pricing Model incorporates an intercept and the betas are not the same as those computed ex ante. The results are illustrated with an example.

## INTRODUCTION

Portfolio selection introduced by Markowitz [1] has many supporters and many detractors. Broadly, the former are those who use his methods successfully and the latter are those who do not. Since its introduction, traditional portfolio selection has undergone much refinement and development. Nonetheless, many of these developments are very similar to or essentially identical to the original method. That is, a portfolio selector remains on a meanvariance efficient frontier. The original theory assumes a quadratic utility function or that the multivariate probability distribution of asset returns is characterized by expected returns and the covariance matrix. Stein's Lemma, Stein [2], and its modern extensions (Liu, [3]; Landsman and Nešlehová, [4]) mean that these remarks are valid under a range of elliptically symmetric distributions and, subject to regularity conditions, for all utility functions. Thus, the efficient frontier should be a robust place to be.

In the previous paragraph, the phrase "a mean-variance efficient frontier" is used deliberately to remind that in practice all efficient frontiers are based on estimates of the underlying parameters, the vector of expected returns and the covariance matrix. Even when consistent estimators of the underlying parameters are used, all efficient frontiers are in reality estimated efficient frontiers. It is well known, by both practitioners and academic researchers, that the ex-post performance of an efficient portfolio often differs substantially from that anticipated at the time of construction. The celebrated papers by Best and Grauer [5] and Chopra and Ziemba [6] document that portfolios which are mean-variance efficient ex ante are sensitive to the inputs; that is to the estimators that are used. As Adcock [7] reports "even in the situation where the user is equipped with good estimates of the input parameters, the outputs are likely to produce results that are different from those expected. In circumstances where the estimates of the inputs are poor, it is inevitable that ex-post performance will be inferior". The recent paper by Kan and Zhou [8] confirms this. These and other difficulties are documented widely, notably in Michaud [9, 10].

The use of estimated values for the model parameters means that it is desirable, even necessary, to use statistical methods to study the behaviour of portfolios which ex ante are mean variance efficient. There is an early work due to Bawa, Brown and Klein [11]. In many papers, the starting point for the use of statistical methods in conjunction with mean-variance portfolio selection is often the work by Jobson and Korkie [12]. This work, in common with other later papers, is concerned with the maximum Sharpe ratio portfolio. If asset returns are IID normal and the usual sample estimators are used, Jobson and Korkie [12] show how to derive expressions for the expected values and variances of the components of the efficient frontier reported in Merton [13]. This version of the efficient frontier allows short positions; that is only the budget constraint is imposed on the expected utility maximization. The resulting formulae in Merton's paper define the shape of the frontier and ex-ante portfolio expected return and variance. They are often referred to collectively as efficient set mathematics. Gibbons, Ross and Shanken [14] present a test of the mean-variance efficiency of a portfolio. This test, which employs a fundamental property of the efficient frontier, is based on a variant of the market model. Under the IID normal assumptions, it results in Hotelling's $T^2$, which apart from a scaling constant has an F distribution. There are similar tests in Huberman and Kandel [15] and Britten-Jones [16]. More recently, Kan and Smith [17] derive expressions for the joint distribution of the components of the efficient frontier given the standard assumptions. To achieve this, they reparameterise the frontier and consider components which are functions of those in Merton's original representation. The results that they derive depend on the Chi-squared and non-central F distributions. Knight and Satchell [18] derive further extensions, specifically for institutional investors. There are several other related works, notably by Bodnar and Schmid [19-21], Hillier and Satchell [22] and Okhrin and Schmid [23].

Under the assumption that the vector of expected returns and the covariance matrix are known, the ex post or actual return on a portfolio is an affine transformation of the vector of asset returns. If returns follow a multivariate normal distribution or any member of the elliptically symmetric class, the distribution of portfolio returns is a member of the same class. The aim of this paper is to present results for the case where the covariance matrix is known, but the vector of expected returns is an estimate or forecast and is therefore a random vector. To avoid duplication, henceforth such a vector is referred to as a forecast. When the joint distribution of returns and the forecast used for portfolio selection is multivariate normal, it is shown that the distribution of ex-post portfolio returns is an extended quadratic form in normal variables. It is shown that this changes the shape of the efficient frontier and leads to different insights into the maximum Sharpe ratio or market portfolio. The results in this paper substantially extend those reported in Adcock [7, 24, and 25].

The paper is set out as follows. Section 2 contains a summary of traditional efficient set mathematics and the assumptions used. Section 3 present the main results of the paper, namely that ex post returns are distributed as an extended quadratic form. Given that the number of possible specifications for the structure of the covariance matrix of asset returns and forecasts is large, Section 4 presents two examples. In Section 5, there are results which examine the effect of the estimated expected returns or forecasts on the Sharpe ratio, the market portfolio and the Capital Asset Pricing Model. Section 6 contains concluding remarks and a brief discussion of potential developments.

## TRADITIONAL EFFICIENT SET MATHEMATICS

Let R be an n-vector of asset returns, which has the multivariate normal distribution $N(\mu, \Sigma)$. The notation $R_p$ denotes portfolio return and $r_f$ the risk free. The notations $1$, $0_n$ and $0_{mn}$ denote respectively an n-vector of ones, an n-vector of zeros and a $m \times n$ matrix of zeros. Subscripts are generally omitted. It is assumed that the covariance matrix $\Sigma$ is non-singular. Maximising expected utility subject only to the budget constraint in the usual way and recalling Stein's Lemma, the first order conditions for portfolio selection lead to the well-known expression for the portfolio weights.

**Ex Post Efficient Set Mathematics**

$$w = \frac{\Sigma^{-1}1}{1^T\Sigma^{-1}1} + \theta\left\{\Sigma^{-1} - \frac{\Sigma^{-1}11^T\Sigma^{-1}}{1^T\Sigma^{-1}1}\right\}\mu$$
$$= w_0 + \theta w_1; \theta \geq 0.$$

The vector $w_0$ is the minimum variance portfolio and satisfies the budget constraint $1^T w_0 = 1$. The vector $w_1$ is a self-financing portfolio. In general, risk appetite $\theta$ is defined as

$$\theta = -E\{U'(R_p)\}/E\{U''(R_p)\}$$

The expected return and variance of portfolio return, which has a normal distribution given the assumptions, are respectively,

$$\mu_p = \alpha_0 + \theta\alpha_1, \sigma_p^2 = \alpha_2 + \theta^2\alpha_1$$

Where the standard constants are defined as

$$\alpha_0 = \frac{\mu^T\Sigma^{-1}1}{1^T\Sigma^{-1}1},$$

$$\alpha_1 = \mu^T\left\{\Sigma^{-1} - \frac{\Sigma^{-1}11^T\Sigma^{-1}}{1^T\Sigma^{-1}1}\right\}\mu,$$

$$\alpha_2 = \frac{1}{1^T\Sigma^{-1}1}.$$

Note that these definitions of the standard constants differ from those in Merton [13]. They are the same as those used in Kan and Smith [17] and are more suitable for the purposes of this paper. The equation of the efficient frontier is

$$\mu_p - \alpha_0 = \sqrt{\alpha_1}\sqrt{\sigma_p^2 - \alpha_2}$$

The market or maximum Sharpe ratio portfolio arises when $\theta_M = \alpha_2 / (\alpha_0 - r_f) = \alpha_2 / \tilde{\alpha}_0$ as long as $\alpha_0 > r_f$. If $\alpha_0 \leq r_f$ the market portfolio does not exist in any meaningful sense.

## DISTRIBUTION OF PORTFOLIO RETURNS

This section presents the main results of the paper, in which it is shown that when m is replaced by a forecast, denoted by F, portfolio return is distributed as an extended quadratic form in normal variables. It is assumed the 2n-vector $X = \begin{bmatrix} R \\ F \end{bmatrix}$ has a non-singular multivariate normal distribution $N(\tau, \Gamma)$ with $\tau = \begin{bmatrix} \mu \\ \mu + \delta \end{bmatrix}$, $\Gamma = \begin{bmatrix} \Sigma_{RR} & \Sigma_{RF} \\ \Sigma_{FR} & \Sigma_{FF} \end{bmatrix}$ respectively. Non-zero entries in the vector $\delta$ mean that the forecast is biased. It is assumed that the covariance matrix is known. The vector of portfolio weights based upon the forecast F is

$$w = w_0 + \theta D_0 F, \quad D_0 = \left\{ \Sigma^{-1} - \frac{\Sigma^{-1} 1 1^T \Sigma^{-1}}{1^T \Sigma^{-1} 1} \right\}$$

Portfolio return is then $R_p = b^T X + 2^{-1} X^T A X$ with $b = \begin{bmatrix} w_0 \\ 0_n \end{bmatrix}$, $A = \begin{bmatrix} 0_{nn} & \theta D_0 \\ \theta D_0 & 0_{nn} \end{bmatrix}$.

Portfolio return is distributed as an extended quadratic form in normal variables. The properties of these are described in detail in Mathai and Prevost [26]. Relevant results for financial applications are in Appendix B of Adcock et al [27]. Specifically, Corollary 2 of their Theorem 2 leads to the following.

## Proposition 1

Apart from an additive constant, portfolio return $R_p$ is distributed as the weighted sum of independent noncentral Chi-squared variables, each with one degree of freedom, and an independently distributed normal variable. That is

$$R_p = \sum_{j=1}^{2(n-1)} \lambda_j \chi^2_{(1),j}\left(\omega_j\right) + \omega_0 + \sigma_0 Z,$$

Where $\lambda_j$ are the $2(n-1)$ non-zero eigenvalues of the matrix $A\Gamma$, $\sigma_0$ and $\omega_j$, $j = 0,1,\dots,2(n-1)$ are scalar functions of elements of the vector $\tau$ and the eigenvectors of $A\Gamma$ and $Z$ is a standard normal variable.

As further technical details of this result are not required for the material that follows below, they are omitted. Briefly, it may be noted that the probability density function of $R_p$ is intractable, although the central limit theorem means that, ceteris paribus, the distribution of $R_p$ will tend to normality as the number of assets increases. This provides support to a finding of Tu and Zhou [28] who suggests that the normality assumption works for the evaluation of portfolio performance. The characteristic function of the extended quadratic form, however, may be inverted numerically using a procedure due to Imhof [29]. Mathai and Prevost [26] note that this procedure may be considered to be exact. The characterristic function is tractable and leads to the following results for the mean and variance of portfolio returns. An outline proof of the following proposition is in Appendix A. It was first reported without proof in Adcock [25].

## Proposition 2

The expected value and variance of portfolio return, denoted with the additional subscript f, are respectively

$$\mu_{pf} = \alpha_0 + \theta\alpha_1 + \theta\beta_0,$$

$$\sigma_{pf}^2 = \alpha_2 + \theta^2\alpha_1 + 2\theta\beta_1 + \theta^2\beta_2,$$

Where $\beta_0$ and $\beta_1$ are

$$\beta_0 = \text{trace}\{D_0\Sigma_{RF}\} + \mu^T D_0\delta, \ \beta_1 = \mu^T D_0\Sigma_{FR}w_0,$$

and

$$\beta_2 = \text{trace}\{(D_0\Sigma_{RF})^2 + D_0\Sigma_{FF}D_0\Sigma_{RR}\}$$
$$+ \mu^T D_0\Sigma_{FF}D_0\mu + \delta^T D_0\Sigma_{RR}D_0\delta$$
$$+ 2\delta^T D_0\Sigma_{RR}D_0\mu + 2\mu^T D_0\Sigma_{FR}D_0(\mu + \delta).$$

The covariance between the returns of an arbitrary portfolio with given weights $w_q$ and an efficient portfolio with risk appetite $\theta$ is

$$\text{cov}(R_q, R_p) = w_q^T\{\theta(\Sigma_{RR} + \Sigma_{RF})D_0\mu + \Sigma_{RR}w_0\}$$

Substitution gives the following:

### *Corollary 2.1*

The equation of the efficient frontier is

$$\mu_{pf} = A_0 + (A_1/\sqrt{B_1})\sqrt{\sigma_{pf}^2 - B_0},$$

Where

$$A_0 = \alpha_0 - \beta_1 (\alpha_1 + \beta_0)/(\alpha_1 + \beta_2), \ A_1 = (\alpha_1 + \beta_0),$$
$$B_0 = \alpha_2 - \beta_1^2/(\alpha_1 + \beta_2), \ B_1 = (\alpha_1 + \beta_2).$$

From Proposition 2 and Corollary 2.1, it is clear that the ex-post expected return and variance of an efficient portfolio constructed using estimates or forecasts of expected returns are different from those based on standard efficient set mathematics. The effect on the maximum Sharpe ratio portfolio is described in Section 5. The detailed effects on mean and variance, and hence the shape of the efficient frontier, depend on the constants $\beta_{0,1,2}$. These in turn depend on $\delta$, the bias in the estimates, and the structure of the covariance matrix $\Gamma$. To illustrate the effects, two examples are presented in Section 4.

## TWO EXAMPLES

The matrices $\Sigma_{RR}$ and $\Sigma_{FF}$ may be written as

$$\Sigma_{RR} = HH^{\mathrm{T}}, \Sigma_{FF} = KK^{\mathrm{T}},$$

Where H and K are full rank $n \times n$ matrices. The covariance matrix of returns and forecasts is

$$\Gamma = \begin{bmatrix} \Sigma_{RR} & HPK^{\mathrm{T}} \\ KP^{\mathrm{T}}H^{\mathrm{T}} & \Sigma_{FF} \end{bmatrix},$$

Where P is the $n \times n$ matrix of cross-correlations between returns and forecasts. It is convenient to define the following scalars

$$\gamma_0 = 1^{\mathrm{T}} H^{-1} 1, \gamma_1 = 1^{\mathrm{T}} H^{-1} w_1, \gamma_2 = 1^{\mathrm{T}} \Sigma_{RR}^{-1} 1.$$

In the two examples below, it is assumed that the covariance matrix of the estimates or forecasts is proportional to $\Sigma_{RR}$, the covariance matrix of asset returns. This is loosely equivalent to assuming that the vector of forecasts is based on simple time's series methods. It is also assumed that forecasts are unbiased $\delta = 0$.

### Rank One Cross-Correlation Matrix with Equal Correlations and Unbiased Estimates

In this case $P = \rho 1\, 1^{\mathrm{T}}$ which leads to

$$
\Gamma = \begin{bmatrix} \Sigma_{RR} & \rho\sqrt{\kappa}\,H 1 1^{\mathrm{T}} K^{\mathrm{T}} \\ \rho\sqrt{\kappa}\,K 1 1^{\mathrm{T}} H^{\mathrm{T}} & \kappa\Sigma_{RR} \end{bmatrix}.
$$

The constants $\beta_{0,1,2}$ are

$$
\beta_0 = \rho\left(n - \gamma_0^2/\gamma_2\right)\sqrt{\kappa}, \quad \beta_1 = \rho\left(\gamma_0\gamma_1/\gamma_2\right)\sqrt{\kappa},
$$
$$
\beta_2 = \kappa\left\{\alpha_1 + (n-1) + \rho^2\left(n - \gamma_0^2/\gamma_2\right)\right\} + 2\rho\gamma_1^2\sqrt{\kappa}.
$$

These are affected by the covariance matrix of asset returns through their dependence on H. Note that 1) by the Cauchy Schwarz inequality $n - \gamma_0^2/\gamma_2 \geq 0$, expected return is increased (decreased) if $\rho$ is positive (negative) and 2) that the requirement that $\Gamma$ be positive semi definite imposes a restriction on $|\rho|$.

## Diagonal Cross-Correlation Matrix with Equal Correlations and Unbiased Estimates

This example was first reported in conference proceedings in Adcock [24]. In this case $P = \rho I$, where $I$ is the $n \times n$ unit matrix, in which case

$$\Gamma = \begin{bmatrix} \Sigma_{RR} & \rho\sqrt{\kappa}\Sigma_{RR} \\ \rho\sqrt{\kappa}\Sigma_{RR} & \kappa\Sigma_{RR} \end{bmatrix},$$

Leading to

$$\beta_0 = \rho(n-1)\sqrt{\kappa}, \quad \beta_1 = 0,$$
$$\beta_2 = \kappa\left\{\alpha_1 + (n-1)\left(1+\rho^2\right)\right\} + 2\rho\alpha_1\sqrt{\kappa}.$$

A special case of this is the use of the sample mean returns based on a time series of length T. In this case $k = T^{-1}$ and $\rho = 0$. There is no effect on mean return, but there is an increase in variance. In particular, the variance is an increasing function of the number of assets.

To illustrate these results a data set consisting of weekly returns from 13 FTSE indices is used. The forecast of the mean returns and the co-variance matrix used are shown in Tables A1 and A2 of Appendix C. The illustration considers five values of correlation r = −0.05, −0.01, 0, 0.01, 0.05. The parameter set for k corresponds to sample sizes of T=1, 5, 10, 50, 100, 1000. The value T=1 may be interpreted as meaning that the covariance matrix associated with the forecasts is predictive, which corresponds with a sensible practice. The $\alpha_{0,1,2}$ and $\beta_{0,1,2}$ are computed using the formulae above. These are shown in Table 1. Panel 1) of Table 1 shows the standard constants. Panel 2) shows the computed values of $\beta_{0,1,2}$ corresponding to values of $\rho$ from −0.05 to 0.05. Note that the values of $\beta_2$ are two orders of magnitude greater than those for $\beta_0$. In panel 3) the column entitled mult0 shows the multiplier to be applied to the Standard Sharpe ratio. Note that for $\rho \geq 0$ the maximum Sharpe ratio occurs at a lower level of risk than the standard case, but that for $\rho < 0$ the maximum Sharpe ratio portfolio is the minimum variance portfolio (MVP). A graph of the efficient frontier for 3 values of $\rho$, namely −0.01, 0.01 and 0.05 and for k=1 is shown in Figure 1. The figure also includes a graph of the conventional efficient frontier. As the figure shows, when $\rho$ is less than zero the efficient frontier is downwards sloping: more risk leads to lower expected return.

**Table 1:** Parameters of the efficient frontier

| 1) Standard Constants | | |
|---|---|---|
| $\alpha_0$ | $\alpha_1$ | $\alpha_2$ |
| 0.0009 | 0.0057 | 0.0004 |

| 2) New Components | | | |
|---|---|---|---|
| $\rho$ | $\beta_0$ | $\beta_1$ | $\beta_2$ |
| -0.05 | -0.6000 | 0.0000 | 12.0351 |
| -0.01 | -0.1200 | 0.0000 | 12.0068 |
| 0 | 0.0000 | 0.0000 | 12.0057 |
| 0.01 | 0.1200 | 0.0000 | 12.0070 |
| 0.05 | 0.6000 | 0.0000 | 12.0363 |

| 3) Parameters of Efficient Frontier | | | |
|---|---|---|---|
| $\rho$ | $\alpha_1 + \beta_0$ | $\alpha_1 + \beta_2$ | mult0 |
| -0.05 | -0.5943 | 12.0408 | -0.0494 |
| -0.01 | -0.1143 | 12.0125 | -0.0095 |
| 0 | 0.0057 | 12.0114 | 0.0005 |
| 0.01 | 0.1257 | 12.0127 | 0.0105 |
| 0.05 | 0.6057 | 12.0420 | 0.0503 |

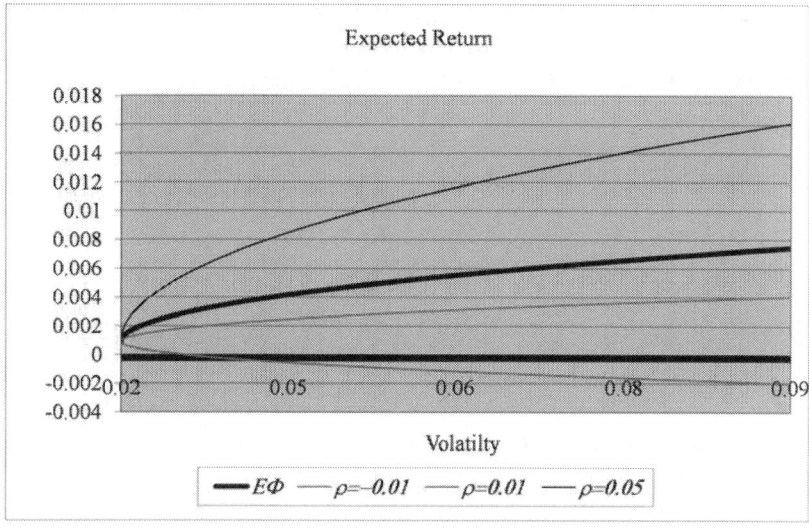

**Figure 1:** The efficient frontier based on forecasts.

For $\rho = 0.01$ the frontier is upwards sloping, but the gradient is always less than that for the ex-ante efficient frontier. When $\rho = 0.05$, the gradient is higher. That is, this value of correlation provides a sufficient signal to outperform the ex-ante frontier. To avoid cluttering the figure other values of $\rho$ are omitted. However, as $\rho$ increases so does the gradient of the frontier. Conversely as $\rho$ decreases from zero, the negative trade-off between risk and expected return becomes progressively worse.

The case $\rho = 0$ may be interpreted as the use of sample returns as a forecast is also omitted. In this case, the corresponding efficient frontier is effectively flat. Figure 2, which is in Section 5, shows the Sharpe ratios plotted against risk appetite $\theta$ for the same values of $\rho$ and for the ex-ante case.

## THE SHARPE RATIO AND THE MARKET PORTFOLIO

In standard efficient set mathematics, the Sharpe ratio is
$SR = (\tilde{\alpha}_0 + \alpha_1\theta) / \sqrt{\alpha_2 + \alpha_1\theta^2}$.

For $\tilde{\alpha}_0 > 0$ the maximum Sharpe ratio or market portfolio is given by
$\theta_M = \alpha_2 / \tilde{\alpha}_0$. Sections 5.1 and 5.2 consider the Sharpe ratio and the
market portfolio for the case when unbiased forecasts of $\mu$ are used,
that is $\delta = 0$. Section 5.3 considers the distribution of returns on the
conventional maximum Sharpe ratio portfolio for the case where the
estimate F is used in place of $\mu$.

### Properties of the Sharpe Ratio

Table 2 shows the differences between the standard Sharpe ratio and its
properties compared with those when the effects of forecasts are taken
into account. Specifically, in the rows labelled 1) the table shows the value
of $\theta$ that maximises the Sharpe ratio, in 2) the value of the Sharpe ratio at
the maximum and in 3) the limiting value as $\theta \to \infty$. The corresponding
results for the case of biased forecasts are substantially more complicated
and so are omitted, but are available on request. In the standard case, a
necessary and sufficient condition for the maximum Sharpe ratio to exist
in a meaningful sense is that $\tilde{\alpha}_0 > 0$. For the expost Sharpe ratio the corresponding condition is $\theta_{Mf} = (\alpha_1 + \beta_0)\tilde{\alpha}_0 / (\alpha_1 + \beta_2) > 0$.

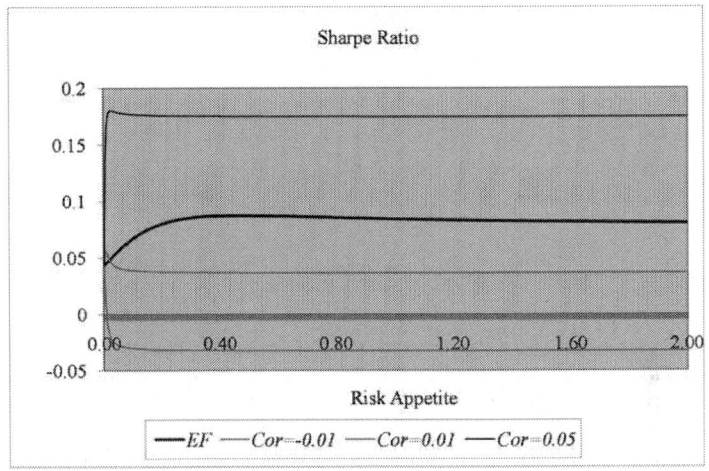

**Figure 2:** Sharpe ratios based on forecasts.

**Table 2:** Comparison of the sharpe ratio

| **Standard** |
|:---:|
| (1) $\theta_M = \alpha_2/\tilde{\alpha}_0$ |
| (2) $SR_M = \sqrt{\alpha_1 + \tilde{\alpha}_0^2/\alpha_2}$ |
| (3) $SR_\infty = \sqrt{\alpha_1}$ |
| **Forecasts** |
| (1) $\theta_{Mf} = \left(\alpha_1 + \beta_0\right)\theta_M/\left(\alpha_1 + \beta_2\right)$ |
| (2) $SR_M = \sqrt{\left(\alpha_1 + \beta_0\right)^2/\left(\alpha_1 + \beta_2\right) + \tilde{\alpha}_0^2/\alpha_2}$ |
| (3) $SR_{\infty f} = \left(\alpha_1 + \beta_0\right)/\sqrt{\alpha_1 + \beta_2}$ |

Figure 2 shows examples of the Sharpe ratio for r = −0.01, 0.01 and 0.05 and for k=1. The standard Sharpe ratio is also shown. When r = −0.01 the maximum Sharpe ratio occurs at the MVP and the ratio declines monotonically as risk increases. For r = 0.01 the maximum is close to the MVP and the Sharpe ratio is always inferior to the ex-ante case. When r = 0.05, however, the Sharpe ratio is superior to the standard case, but the maximum is attained at lower risk.

## The Market Portfolio and the CAPM

The question of the market portfolio under forecast uncertainty naturally arises. Standard manipulations lead to the following.

### Proposition 3

Given the assumptions above, the maximum ex post Sharpe ratio portfolio is the ex post market portfolio.

Thus, although the ex post market portfolio differs from that found ex ante, there is a corresponding capital market line, whose intercept is the risk free rate. The argument that investors will hold a combination of lending/borrowing and the market portfolio still holds. This is subject to the assumption that the joint distribution of returns and forecasts is the same for all market participants. This result leads in turn to the question of the CAPM. Under the assumptions of the paper, is the expected excess return on an asset or portfolio given by the product of beta and the expected excess return on the market portfolio? The treatment below follows that in Chapter 3 of Huang and Litzenberger [30] and requires the following result.

### Proposition 4

The covariance of two efficient portfolios p and q is

$$\text{cov}\left(R_p, R_q\right) = \alpha_2 + \theta_p \theta_q \left(\alpha_1 + \beta_2\right) + \left(\theta_p + \theta_q\right)\beta_1 .$$

This leads to portfolio q being a zero-beta portfolio with respect to portfolio p if its risk appetite is

$$\theta_z = -\left(\alpha_2 + \beta_1\theta_p\right)/\left\{\beta_1 + \left(\alpha_1 + \beta_2\right)\theta_p\right\}.$$

Note that for the case where forecasts are biased $\beta_1 \neq 0$ and portfolio p may be any portfolio including the minimum variance portfolio. For the case considered in detail in this section, $\beta_1 = 0$ and portfolio p can be any portfolio except the minimum variance portfolio. For this case, the expected return on the zero beta portfolio is

$$E\left(R_z\right) = \alpha_0 - \alpha_2\left(\alpha_1 + \beta_0\right)/\left(\alpha_1 + \beta_2\right)\theta_p$$

Standard manipulations, similar to those in Chapter 3 of Huang and Litzenberger, lead to the following

## *Proposition 5*

The intercept of the straight line that is the tangent to the efficient frontier at portfolio p is equal to $E(R_Z)$.

## *Proposition 6*

If portfolio p is the market portfolio, the expected return on the zero beta portfolio equals the risk free rate $r_f$.

In the standard case where $\mu$ is given, consideration of the covariance between the returns of portfolio p and an arbitrary portfolio leads to the CAPM if portfolio p is in fact the market portfolio. For the case considered in this paper, Proposition 2 leads to a modified version of the CAPM

## Proposition 7

Let q be any portfolio with weights $w_q$, M be the ex post market portfolio and let $\tilde{\mu}_q$ and $\tilde{\mu}_M$ be their respective expected excess returns. When $\delta = 0$, it follows that $\tilde{\mu}_q = A + B\tilde{\mu}_M$ where B is the beta of portfolio q with respect to M

$$B = \Psi_0 + \Psi_1 \left[ \left\{ \tilde{\mu}_q + \theta_M \left( \beta_0 - \beta_2 \right) + w_q^T \Sigma_{RF} w_1 \right\} \big/ \tilde{\mu}_M \right],$$

Where

$$\Psi_0 = \left( \beta_2 - \beta_0 \right) \big/ \left( \alpha_1 + \beta_2 \right), \Psi_1 = \left( \alpha_1 + \beta_0 \right) \big/ \left( \alpha_1 + \beta_2 \right).$$

Note that this reduces to the standard case when $\mu$ is given, but that for this case the intercept is not zero in general.

Continuing the example, Table 3 contains values of alpha and beta for two portfolios for the values of $\rho$ and k used above. The first portfolio is an equally weighted portfolio of returns on the 13 FTSE indices. The second is the conventional market portfolio for which the weights are proportional to $\Sigma_{RR}^{-1}\tilde{\mu}$. Panel 1) of Table 3 shows the alphas and betas for the equally weighted portfolio. These are computed for the standard efficient frontier (table rows called EF). They are also computed for the specified values of $\rho$ and k. As the table shows, the values of alpha are non-zero. They are numerically small, but of comparable magnitude to the return forecasts shown in appendix Table A1. The values of beta decrease as $\rho$ increases. Both alpha and beta approach their standard values as k decreases to zero, equivalently the implicit sample size increases. Similar behaviour is observed in Panel 2), although it is notable that the alphas are substantial when $|\rho| = 0.05$. It is also notable that beta is a non-linear function of both $\rho$ and k, with the phenomenon being more apparent for the conventional market portfolio.

## Property of the Maximum Sharpe Ratio Portfolio

When F is used as the forecast of expected return, the maximum Sharpe ratio portfolio has weights given by

$$w_M = \Sigma_{RR}^{-1} F / 1^T \Sigma_{RR}^{-1} F .$$

The return on the market portfolio is

$$R_{Mf} = R^T \Sigma_{RR}^{-1} F / 1^T \Sigma_{RR}^{-1} F .$$

The following interesting result is proved in Appendix B.

### Proposition 8

Given the assumptions above, the expected value of the market portfolio based on forecast of the expected return is undefined.

Strictly speaking, the result is of theoretical interest. Nonetheless, it suggests that returns on the maximum Sharpe ratio portfolio based on estimates may in practice be volatile.

## DISCUSSION AND CONCLUDING REMARKS

This paper considers efficient set mathematics for the case where the covariance matrix of asset returns is assumed to be known but ex ante the vector of expected returns is replaced by an estimated or forecast value. It is shown that the ex post mean and variance differ from the standard results. Consequently the maximum Sharpe ratio portfolio also differs from the standard result. This portfolio remains the market portfolio. Thus, even with uncertainty about the vector of expected returns, subject to the assumptions made about the joint distribution of actual returns and estimated mean returns, ex post Sharpe ratio maximizers hold the ex post market portfolio.

**Table 3:** Behaviour of alpha and beta

| | Sample Size Equivalent | | | | |
|---|---|---|---|---|---|
| | 1 | 10 | 50 | 500 | +1e6 |
| | (1) Equally weighted portfolio | | | | |
| | Alphaq0 | | | | |
| EF | 0.0000 | 0.0000 | 0.0000 | 0.0000 | 0.0000 |
| -0.05 | 0.0004 | 0.0004 | 0.0005 | 0.0006 | 0.0000 |
| -0.01 | 0.0004 | 0.0004 | 0.0004 | 0.0004 | 0.0000 |
| 0 | 0.0004 | 0.0004 | 0.0004 | 0.0003 | 0.0000 |
| 0.01 | 0.0004 | 0.0004 | 0.0003 | 0.0002 | 0.0000 |
| 0.05 | 0.0003 | 0.0003 | 0.0002 | 0.0000 | 0.0000 |
| | Bataq0 | | | | |
| EF | 0.3520 | 0.3520 | 0.3520 | 0.3520 | 0.3520 |
| -0.05 | 0.0590 | 0.0589 | 0.0595 | 0.0780 | 0.4016 |
| -0.01 | 0.6307 | 0.6776 | 0.7690 | 1.0027 | 0.3615 |
| 0 | 0.9988 | 0.9878 | 0.9434 | 0.6832 | 0.3523 |
| 0.01 | 0.5907 | 0.5515 | 0.4916 | 0.3621 | 0.3435 |
| 0.05 | 0.0593 | 0.0600 | 0.0617 | 0.0733 | 0.3114 |
| | (2) Standard Market Portfolio | | | | |
| | Alpham0 | | | | |
| EF | 0.0000 | 0.0000 | 0.0000 | 0.0000 | 0.0000 |
| -0.05 | 0.0142 | 0.0149 | 0.0155 | 0.0119 | 0.0002 |
| -0.01 | 0.0005 | 0.0004 | 0.0002 | 0.0000 | 0.0001 |
| 0 | 0.0000 | 0.0000 | 0.0001 | 0.0004 | 0.0000 |
| 0.01 | 0.0006 | 0.0007 | 0.0009 | 0.0012 | -0.0001 |
| 0.05 | 0.0134 | 0.0122 | 0.0097 | -0.0015 | -0.0003 |
| | Betam0 | | | | |
| EF | 1.0000 | 1.0000 | 1.0000 | 1.0000 | 1.0000 |
| -0.05 | 0.0142 | -0.0794 | -0.2300 | -0.5201 | 0.9241 |
| -0.01 | 0.6299 | 0.6771 | 0.7743 | 0.9981 | 0.9828 |
| 0 | 0.9986 | 0.9859 | 0.9358 | 0.7028 | 0.9984 |
| 0.01 | 0.5926 | 0.5599 | 0.5220 | 0.5690 | 1.0144 |
| 0.05 | 0.1059 | 0.2091 | 0.4016 | 1.0876 | 1.0810 |

The properties of the zero beta portfolio are also similar to the standard results. A notable exception, however, is that the capital asset pricing model incorporates an intercept and the ex post betas are not the same as those computed ex ante.

The numerical example provides a demonstration of well-known empirical features: positive correlations between returns and estimates improve ex post portfolio performance; negative correlations damage it; volatility ex post may be expected to be higher than that predicted ex ante.

The assumption of multivariate normality with known covariance matrix is a limitation of the results, except perhaps for those of low frequency. The results presented here imply that a tractable model for the multivariate probability distribution of returns and estimates is required. Scale mixtures of the multivariate normal distribution are an obvious candidate. The use of multivariate distributions which incorporate skewness is an open research question.

## APPENDICES

### A—Moments of Extended Quadratic Forms in Normal Variables
In the notation of section 3, portfolio return is

$$R_p = b^T X + 2^{-1} X^T A X$$

This may be written as the quadratic form

$$R_p = 2^{-1} \tilde{X}^T \tilde{A} \tilde{X} ,$$

Where

$$\tilde{X} = \begin{bmatrix} X \\ b \end{bmatrix}, \tilde{A} = \begin{bmatrix} A & I \\ I & 0 \end{bmatrix}.$$

The vector $\tilde{X}$ has a singular multivariate normal distribution with mean vector and covariance matrix respectively.

$$\tilde{\tau} = \begin{bmatrix} \tau \\ b \end{bmatrix}, \tilde{\Gamma} = \begin{bmatrix} \Gamma & 0 \\ 0 & 0 \end{bmatrix},$$

For a random vector $Y$ which has the general multivariate normal distribution $N(\mu, \Sigma)$ standard results are that the cumulants of the quadratic form $Y^T B Y$ are

$$\kappa_k = 2^{k-1}(k-1)!\left[ \operatorname{trace}\left\{ (\Sigma B)^k \right\} + k\mu^T B^{(k)} \mu \right],$$

Where $B^{(1)} = B, B^{(2)} = B\Sigma B$ and so on, and that

$$\operatorname{cov}\left( d^T Y, Y^T B Y \right) = d^T \Sigma B \mu .$$

Substitution of $\tilde{\tau}, \tilde{\Gamma}, \tilde{A}$ and $w_q$ gives the results of Proposition 2.

## B — Proof of Proposition 8

The return on the market portfolio is $R_{Mf} = R^T \Sigma_{RR}^{-1} F / 1^T \Sigma_{RR}^{-1} F$, where $R$ and $F$ have the multivariate normal distribution $N(\tau, \Gamma)$ as defined in Section 3. Conditional on $F=f$, the expected value of $R_{Mf}$ is

$$\left[ f^T \left\{ \Sigma_{RR}^{-1}\mu - \Sigma_{RF}^{-1}(\mu + \delta) \right\} + f^T \Sigma_{RF}^{-1} f \right] / f^T \Sigma_{RF}^{-1} 1 .$$

The first term is the ratio of two variables which have a bivariate normal distribution. Cedilnik, Košmelj and Blejec [31] show that such a variable does not have an expected value or higher moments. This is sufficient to ensure that the unconditional moments of $R_{Mf}$ are undefined.

## C—Forecast Mean Vector and Covariance Matrix

### Table A1: Forecast Mean Weekly Returns for 13 FTSE Indices

| No. | Index | Forecast |
|-----|-------|----------|
| 1 | FTSE100 | 0.0012 |
| 2 | FTSE250 | 0.0017 |
| 3 | FTS250-ex-Inv | 0.0017 |
| 4 | FTSE350 | 0.0013 |
| 5 | FTSE350-ex-Inv | 0.0013 |
| 6 | FTSE350-HY | 0.0013 |
| 7 | FTSE350-LY | 0.0011 |
| 8 | FTSE-SC | 0.0012 |
| 9 | FTSE-Sex-Inv | 0.0010 |
| 10 | FTSE-All-Share | 0.0013 |
| 11 | FTSE-AS-ex-Inv | 0.0013 |
| 12 | FTSE-AS-ex-mult | 0.0011 |
| 13 | FTSE-Aim | 0.0006 |

### Table A2: Sample Covariance/Correlation Matrix

| Index | 1 | 2 | 3 | 4 | 5 |
|-------|------|------|------|------|------|
| 1 | 0.0007 | 0.0006 | 0.0006 | 0.0007 | 0.0006 |
| 2 | 0.8440 | 0.0007 | 0.0008 | 0.0006 | 0.0006 |
| 3 | 0.8271 | 0.9954 | 0.0008 | 0.0006 | 0.0006 |
| 4 | 0.9774 | 0.8649 | 0.8501 | 0.0007 | 0.0006 |
| 5 | 0.9627 | 0.8478 | 0.8346 | 0.9670 | 0.0007 |
| 6 | 0.8848 | 0.7628 | 0.7466 | 0.8698 | 0.8762 |
| 7 | 0.9176 | 0.8430 | 0.8312 | 0.9089 | 0.9271 |
| 8 | 0.6515 | 0.7884 | 0.7809 | 0.6736 | 0.6872 |
| 9 | 0.5913 | 0.7630 | 0.7617 | 0.6174 | 0.5894 |
| 10 | 0.9617 | 0.8493 | 0.8354 | 0.9544 | 0.9791 |

| Index | 1 | 2 | 3 | 4 | 5 |
|---|---|---|---|---|---|
| 11 | 0.9559 | 0.8454 | 0.8323 | 0.9459 | 0.9716 |
| 12 | 0.8747 | 0.8703 | 0.8652 | 0.8739 | 0.8950 |
| 13 | 0.6445 | 0.7265 | 0.7176 | 0.6656 | 0.6458 |

| Index | 6 | 7 | 8 | 9 | 10 |
|---|---|---|---|---|---|
| 1 | 0.0006 | 0.0007 | 0.0004 | 0.0004 | 0.0006 |
| 2 | 0.0005 | 0.0006 | 0.0005 | 0.0005 | 0.0006 |
| 3 | 0.0005 | 0.0007 | 0.0005 | 0.0005 | 0.0006 |
| 4 | 0.0006 | 0.0007 | 0.0004 | 0.0004 | 0.0006 |
| 5 | 0.0006 | 0.0007 | 0.0004 | 0.0004 | 0.0006 |
| 6 | 0.0006 | 0.0005 | 0.0003 | 0.0003 | 0.0006 |
| 7 | 0.7598 | 0.0008 | 0.0004 | 0.0004 | 0.0007 |
| 8 | 0.6093 | 0.6714 | 0.0005 | 0.0005 | 0.0004 |
| 9 | 0.5245 | 0.5918 | 0.8130 | 0.0006 | 0.0004 |
| 10 | 0.8755 | 0.9327 | 0.6966 | 0.5990 | 0.0006 |
| 11 | 0.8667 | 0.9254 | 0.6903 | 0.5934 | 0.9701 |
| 12 | 0.8206 | 0.8403 | 0.7181 | 0.6468 | 0.8943 |
| 13 | 0.5298 | 0.6745 | 0.7548 | 0.7508 | 0.6617 |

| Index | 11 | 12 | 13 |
|---|---|---|---|
| 1 | 0.0006 | 0.0006 | 0.0004 |
| 2 | 0.0006 | 0.0006 | 0.0005 |
| 3 | 0.0006 | 0.0006 | 0.0005 |
| 4 | 0.0006 | 0.0006 | 0.0004 |
| 5 | 0.0006 | 0.0006 | 0.0004 |
| 6 | 0.0006 | 0.0005 | 0.0003 |
| 7 | 0.0007 | 0.0006 | 0.0005 |
| 8 | 0.0004 | 0.0004 | 0.0004 |
| 9 | 0.0004 | 0.0004 | 0.0005 |
| 10 | 0.0006 | 0.0006 | 0.0004 |
| 11 | 0.0007 | 0.0006 | 0.0004 |
| 12 | 0.8896 | 0.0007 | 0.0004 |
| 13 | 0.6472 | 0.6434 | 0.0006 |

Correlations are shown below the leading diagonal.

## REFERENCES

1.  H. Markowitz, "Portfolio Selection," Journal of Finance, Vol. 7, No. 1, 1952, pp. 77–91.

2.  C. M. Stein, "Estimation of the Mean of a Multivariate Normal Distribution," Annals of Statistics, Vol. 9, No. 6, 1981, pp. 1135–1151. doi:10.1214/aos/1176345632.

3.  J. S. Liu, "Siegel's Formula via Stein's Identities," Statistics and Probability Letters, Vol. 21, No. 3, 1994, pp. 247–251. doi:10.1016/0167-7152(94)90121-X.

4.  Z. Landsman and J. Nešlehová, "Stein's Lemma for Elliptical Random Vectors," Journal of Multivariate Analysis, Vol. 99, No. 5, 2008, pp. 912–927. doi:10.1016/j.jmva.2007.05.006.

5.  M. J. Best and R. R. Grauer, "On the Sensitivity of MeanVariance-Efficient Portfolios to Changes in Asset Means: Some Analytical and Computational Results," Review of Financial Studies, Vol. 4, No. 2, 1991, pp. 315–342. doi:10.1093/rfs/4.2.315.

6.  V. Chopra and W. T. Ziemba, "The Effect of Errors in Means, Variances and Covariances on Optimal Portfolio Choice," Journal of Portfolio Management, Vol. 19, No. 2, 1993, pp. 6–11. doi:10.3905/jpm.1993.409440.

7.  C. J. Adcock, "Predicting Portfolio Returns Using The Distributions of Efficient Set Portfolios," In S. E. Satchell and A Scowcroft, Eds., Advances in Portfolio Construction and Implementation, Butterworth Heinemann, Oxford, 2003, pp. 342–355.

8.  R. Kan and G. Zhou, "Optimal Portfolio Choice with Parameter Uncertainty," Journal of Financial and Quantitative Analysis, Vol. 42, No. 3, 2007, pp. 621–656. doi:10.1017/S0022109000004129.

9.  R. O. Michaud, "The Markowitz Optimization Enigma: Is Optimized Optimal?" Financial Analysts Journal, 1989, pp. 31–42.

10. R. O. Michaud, "Efficient Asset Management," Harvard Business School Press, Boston, 1998.

11. V. Bawa, S. J. Brown and R. Klein, "Estimation Risk and Optimal Portfolio Choice," Studies in Bayesian Econometrics, North Holland, Amsterdam, Vol. 3, 1979.

12. J. D. Jobson and B. Korkie, "Estimation for Markowitz Efficient Portfolios," Journal of the American Statistical Association, Vol. 75, No. 371, 1980, pp. 544–554. doi:10.1080/01621459.1980.10477507.

13. R. Merton, "An Analytical Derivation of the Efficient Portfolio Frontier," Journal of Financial and Quantitative Analysis, Vol. 7, No. 4, 1972, pp. 1851–1872. doi:10.2307/2329621.

14. M. R. Gibbons, S. A. Ross and J. Shanken, "A Test of the Efficiency of a Given Portfolio," Econometrica, Vol. 57, No. 5, 1989, pp. 1121–1152. doi:10.2307/1913625.

15. G. Huberman and S. Kandel, "Mean-Variance Spanning," The Journal of Finance, Vol. 42, No. 4, 1987, pp. 873–888. doi:10.1111/j.1540-6261.1987.tb03917.x.

16. M. Britten-Jones, "The Sampling Error in Estimates of Mean-Variance Efficient Portfolio Weights". Journal of Finance, Vol. 54, No. 2, 1999, pp. 655–672. doi:10.1111/0022-1082.00120.

17. R. Kan, and D. R. Smith, "The Distribution of the Sample Minimum-Variance Frontier," Management Science, Vol. 54, No. 7, 2008, pp. 1364–1360.doi:10.1287/mnsc.1070.0852.

18. J. Knight, and S. E. Satchell, "Exact Properties of Measures of Optimal Investment for Benchmarked Portfolios," Quantitative Finance, Vol. 10, No. 5, 2010, pp. 495–502.doi:10.1080/14697680903061412.

19. T. Bodnar, and W. Schmid, "A Test for the Weights of the Global Minimum Variance Portfolio in an Elliptical Model," Metrika, Vol. 67, No. 2, 2008, pp. 127–143.doi:10.1007/s00184-007-0126-7.

20. T. Bodnar and W. Schmid "Estimation of Optimal Portfolio Compositions for Gaussian Returns," Statistics & Decisions, Vol. 26, No. 3, 2008, pp. 179–201. doi:10.1524/stnd.2008.0918.

21. T. Bodnar and W. Schmid "Econometrical Analysis of the Sample Efficient Frontier," The European Journal of Finance, Vol. 15, No. 3, 2009, pp. 317–335. doi:10.1080/13518470802423478.

22. G. H. Hillier and S. E. Satchell, "Some Exact Results for Efficient Portfolios with Given Returns," In S. E. Satchell and A Scowcroft, Eds., Advances in Portfolio Construction and Implementation, Butterworth Heinemann, Oxford, 2003, pp. 310–325.

23. Y. Okhrin and W. Schmid, "Distributional Properties of Portfolio Weights," Journal of Econometrics, Vol. 134, No. 1, 2006, pp. 235–256. doi:10.1016/j.jeconom.2005.06.022.

24. C. J. Adcock, "The Statistical Properties of Optimised Portfolios," Proceedings of the 1996 Chemical Bank— Imperial College Conference on Forecasting Financial Markets, London, 1996.

25. C. J. Adcock, "Dynamic Control of Risk in Optimised Portfolios," The IMA Journal of Mathematics Applied in Business and Industry, Vol. 11, No. 1, 2000, pp. 27–138.

26. M. Mathai and S. B. Prevost, "Quadratic Forms in Random Variables," Springer, Heidelberg, 1992.

27. C. J. Adcock, M. C. Cortez, M. R. Armada and F. Silva "Time Varying Betas and the Unconditional Distribution of Asset Returns," Quantitative Finance, Vol. 12, No. 6, 2012, pp. 951–967. doi:10.1080/14697688.2010.544667.

28. J. Tu and G. Zhou, "Data-Generating Process Uncertainty: What Difference Does It Make in Portfolio Decisions?" Journal of Financial Economics, Vol. 72, No. 2, 2004, pp. 385–421.doi:10.1016/j.jfineco.2003.05.003.

29. J. P. Imhof, "Computing the Distribution of Quadratic Forms in Normal Variables," Biometrika, Vol. 48, No. 3, 1961, pp. 419–426.

30. C.-F. Huang and R. H. Litzenberger, "Foundations for Financial Economics," Prentice Hall, Englewood Cliffs, 1988.

31. Cedilnik, K Košmelj and A. Blejec, "The Distribution of the Ratio of Jointly Normal Variables," Metodološki Zvezki, Vol. 1, No. 1, 2004, pp. 99–108.

## CITATION

C. Adcock, "Ex Post Efficient Set Mathematics," Journal of Mathematical Finance, Vol. 3 No. 1A, 2013, pp. 201-210. doi: 10.4236/jmf.2013.31A019.

# Low-Rank Positive Approximants of Symmetric Matrices

## *Achiya Dax*
Hydrological Service, Jerusalem, Israel

## ABSTRACT

Given a symmetric matrix X, we consider the problem of finding a low-rank positive approximant of X. That is, a symmetric positive semi definite matrix, S, whose rank is smaller than a given positive integer, l, which is nearest to X in a certain matrix norm. The problem is first solved with regard to four common norms: The Frobenius norm, the Schatten p-norm, the trace norm, and the spectral norm. Then the solution is extended to any unitarily invariant matrix norm. The proof is based on a subtle combination of Ky Fan dominance theorem, a modified pinching principle, and Mirsky minimum-norm theorem.

## INTRODUCTION

Let X be a given real symmetric n×n matrix. In this paper we consider the problem of finding a low-rank symmetric positive semidefinite matrix which is nearest to X with regard to a certain matrix norm. Let $\|\cdot\|$ be a given unitarily invariant matrix norm on $\mathbb{R}^{n \times n}$. (The basic features of such norms are explained in the next section.) Let l be a given positive integer such that $1 \le l \le n - 1$, and define

$$\mathbb{S}^+_{n,\ell} = \{S \mid S \in \mathbb{R}^{n\times n}, S \geq 0, \text{ and } \operatorname{rank}(S) \leq \ell\},$$

where the notation $S \geq 0$ means that S is symmetric and positive semi definite. Then the problem to solve has the form

minimize    $F(S) = \|X - S\|$

subject to    $S \in \mathbb{S}^+_{n,\ell}.$

(1.1)

The need for solving such problems arises in certain matrix completion methods that consider Euclidean distance matrices, see [1] or [2]. Since X is assumed to be a symmetric matrix, it has a spectral decomposition

$$X = Q\Lambda Q^{\mathrm{T}},$$

(1.2)

Where $Q \in \mathbb{R}^{n\times n}, Q^{\mathrm{T}}Q = I$, is an orthonormal matrix.

$$\Lambda = \operatorname{diag}\{\lambda_1, \lambda_2, \cdots, \lambda_n\}$$

is a diagonal matrix, and

$$\lambda_1 \geq \lambda_2 \geq \cdots \geq \lambda_n$$

are the eigenvalues of X in decreasing order. If $X \geq 0$, then $\lambda_n \geq 0$ and the spectral decomposition (1.2) coincides with the SVD of X. In this case the solution of (1.1) is given by the Eckart-Young-Mirsky theorem. See the next section.

The rest of the paper assumes, therefore, that $\lambda_n < 0$. In this case the solution of (1.1) is related to that of the problem

$$\text{minimize} \quad F(S) = \|X - S\|$$
$$\text{subject to} \quad S \in \mathbb{S}_n^+, \tag{1.3}$$

where

$$\mathbb{S}_n^+ = \left\{ S \mid S \in \mathbb{R}^{n \times n} \text{ and } S \geq 0 \right\}.$$

Let q be a non-negative integer that counts the number of positive eigenvalues. That is

$$\lambda_j > 0 \text{ for } j = 1, \cdots, q, \text{ and } \lambda_j \leq 0 \text{ for } j = q+1, \cdots, n. \tag{1.4}$$

Let the diagonal matrix $\Lambda_q$ denotes the positive part of $\Lambda$,

$$\Lambda_q = \text{diag}\left\{ \lambda_1, \cdots, \lambda_q, 0, \cdots, 0 \right\}. \tag{1.5}$$

(If $\lambda_1 \leq 0$ then q=0, and $\Lambda_0$ is the null matrix.) Then, as we shall see in Section 3, the matrix

$$X_q = Q\Lambda_q Q^T \in \mathbb{S}_n^+$$

solves (1.3) in any unitarily invariant norm.

If $q \leq l$ then, clearly, $X_q$ is also a solution of (1.1). Hence in the rest of the paper we assume that $1 \leq l < q$. This assumption implies that the diagonal matrix

$$\Lambda_\ell = \text{diag}\left\{ \lambda_1, \cdots, \lambda_\ell, 0, \cdots, 0 \right\} \tag{1.6}$$

belongs to $\mathbb{S}_{n,l}^{+}$. The aim of the paper is to show that the matrix

$$X_{c} = Q\Lambda_{t}Q^{T}$$

solves (1.1) for any unitarily invariant norm.

Let $A \in \mathbb{R}^{n \times n}$ be a given real $n \times n$ matrix. Then another related problem is

minimize $\quad F(S) = \|A - S\|$

subject to $\quad S \in \mathbb{S}_{n}^{+},$

$$(1.7)$$

The relation between (1.7) and (1.3) is seen when using the Frobenius matrix norm. Let $S \in \mathbb{R}^{n \times n}$ be a symmetric matrix and let $T \in \mathbb{R}^{n \times n}$ be a skew-symmetric matrix. Then, clearly,

$$\|S + T\|_{F}^{2} = \|S\|_{F}^{2} + \|T\|_{F}^{2}.$$

$$(1.8)$$

Recall also that any matrix $A \in \mathbb{R}^{n \times n}$ has a unique presentation as the sum A=X+Y where $X = (A^{T} + A)/2$ is symmetric and $Y = (A^{T} + A)/2$ is skew-symmetric. Consequently, for any symmetric matrix, $S$ say,

$$\|A - S\|_{F}^{2} = \|X - S + Y\|_{F}^{2} = \|X - S\|_{F}^{2} + \|Y\|_{F}^{2}.$$

$$(1.9)$$

Therefore, when using the Frobenius norm, a solution of (1.3) provides a solution of (1.7). This observation is due to Higham [3]. A matrix that solves (1.7) or (1.3) is called "positive approximant". Similarly, the term "low-rank positive approximant" refers to a matrix that solves (1.1).

The current interest in positive approximants was initiated in Halmos' paper [4], which considers the solution of (1.7) in the spectral norm. Rogers and Ward [5] considered the solution of (1.7) in the Schatten-p norm, Ando [6] considered this problem in the trace norm, and Higham [3] considered the Frobenius norm. Halmos [4] has considered the positive approximant problem in a more general context of linear operators on a Hilbert space. Other positive approximants problems (in the operators context) are considered in [7] - [11]. The problems (1.1), (1.3) and (1.7) fall into the category of "matrix nearness problems". Further examples of matrix (or operator) nearness problems are discussed in [12] - [18]. A review of this topic is given in Higham [19].

The plan of the paper is as follows. In the next section we introduce notations and tools which are needed for the coming discussions. In Section 3 we show that $X_q$ solves (1.3). Section 4 considers the solution of (1.1) in Frobenius norm. This involves the Eckart-Young theorem. In the next sections Mirsky theorem extends the results to Schatten-p norms, the trace norm, and the spectral norm. Then it is proved that $X_l$ solves (1.1) in any unitarily invariant norm. The proof of this claim requires a subtle combination of Ky Fan dominance theorem, a modified pinching principle, and Mirsky theorem.

## NOTATIONS AND TOOLS

In this section we introduce notations and facts which are needed for coming discussions. Here A denotes a real $m \times n$ matrix with $m \geq n$. Let

$$A = USV^{\mathrm{T}} \tag{2.1}$$

be an SVD of A, where $U = [u_1, \cdots, u_m]$ is an $m \times m$ orthogonal matrix, $V = [v_1, \cdots, v_n]$ is an $n \times n$ orthogonal matrix, and $S = \mathrm{diag}\{\sigma_1, \cdots, \sigma_n\}$ is an $m \times n$ diagonal matrix. The singular values of A are assumed to be nonnegative and sorted to satisfy

$$\sigma_1 \geq \sigma_2 \geq \cdots \geq \sigma_n \geq 0. \tag{2.2}$$

The columns of U and V are called left singular vectors and right singular vectors, respectively. These vectors are related by the equalities

$$A\mathbf{v}_j = \sigma_j \mathbf{u}_j \quad \text{and} \quad A^{\mathrm{T}}\mathbf{u}_j = \sigma_j \mathbf{v}_j, \quad j = 1, \cdots, n. \tag{2.3}$$

A further consequence of (2.1) is the equality

$$A = \sum_{j=1}^{n} \sigma_j \mathbf{u}_j \mathbf{v}_j^{\mathrm{T}}. \tag{2.4}$$

Moreover, let r denotes the rank of A. Then, clearly,

$$\sigma_1 \geq \cdots \geq \sigma_r > 0 \quad \text{and} \quad \sigma_j = 0 \quad \text{for } j = r+1, \cdots, n. \tag{2.5}$$

So (2.4) can be rewritten as

$$A = \sum_{j=1}^{r} \sigma_j \mathbf{u}_j \mathbf{v}_j^{\mathrm{T}}. \tag{2.6}$$

Let the matrices

$$U_k = [\mathbf{u}_1, \cdots, \mathbf{u}_k] \in \mathbb{R}^{m \times k} \quad \text{and} \quad V_k = [\mathbf{v}_1, \cdots, \mathbf{v}_k] \in \mathbb{R}^{n \times k} \tag{2.7}$$

be constructed from the first k columns of U and V, respectively. Let $S_k = diag\{\sigma_1, \cdots, \sigma_k\}$ be a $k \times k$ diagonal matrix. Then the matrix

$$T_k = U_k S_k V_k^{\mathrm{T}} = \sum_{j=1}^{k} \sigma_j \mathbf{u}_j \mathbf{v}_j^{\mathrm{T}} \tag{2.8}$$

is called a rank-k truncated SVD of A. (If $\sigma_k = \sigma_{k+1}$ then this matrix is not unique.)

Let $a_{ij}, u_{ij}, v_{ij}$ denote the (i, j) entries of the matrices A, U, V, respectively. Then (2.4) indicates that

$$a_{ii} = \sum_{j=1}^{n} \sigma_j u_{ij} v_{ij} \quad \text{for } i = 1, \cdots, n$$

(2.9)

and

$$\sum_{i=1}^{n} |a_{ii}| \le \sum_{i=1}^{n}\sum_{j=1}^{n} \sigma_j |u_{ij}| \cdot |v_{ij}| = \sum_{j=1}^{n} \sigma_j \sum_{i=1}^{n} |u_{ij}| \cdot |v_{ij}| \le \sum_{j=1}^{n} \sigma_j,$$

(2.10)

where the last inequality follows from the Cauchy-Schwarz inequality and the fact that the columns of U and V have unit length.

Another useful property regards the concepts of majorization and unitarily invariant norms. Recall that a matrix norm $\|\cdot\|$ on $\mathbb{R}^{m\times n}$ is called unitarily invariant if the equalities

$$\|A\| = \|X^{\mathrm{T}} A\| = \|AY\| = \|X^{\mathrm{T}} AY\|$$

(2.11)

are satisfied for any matrix $A \in \mathbb{R}^{m\times n}$, and any pair of unitary matrices $X \in \mathbb{R}^{m\times m}$ and $Y \in \mathbb{R}^{n\times n}$. Let B and C be a given pair of $m\times n$ matrices with singular values

$$\beta_1 \ge \beta_2 \ge \cdots \ge \beta_n \ge 0 \quad \text{and} \quad \gamma_1 \ge \gamma_2 \ge \cdots \ge \gamma_n \ge 0,$$

respectively. Let $\beta = (\beta_1, \cdots, \beta_n)^T$ and $\gamma = (\gamma_1, \cdots, \gamma_n)^T$ denote the corresponding n-vectors of singular values. Then the weak majorization relation $\beta <_\omega \gamma$ means that these vectors satisfy the inequality

$$\sum_{j=1}^{k} \beta_j \leq \sum_{j=1}^{k} \gamma_j \quad \text{for } k = 1, \cdots, n.$$

(2.12)

In this case we say that $\beta$ is weakly majorized by $\gamma$, or that the singular values of B are weakly majorized by those of C. The dominance theorem of Fan [20] [21] relates these two concepts. It says that if the singular values of B are weakly majorized by those of C then the inequality

$$\|B\| \leq \|C\|$$

(2.13)

holds for any unitarily invariant norm. For detailed proof of this fact see, for example, [8], [20] - [23]. The most popular example of an unitarily invariant norm is, perhaps, the Frobenius matrix norm

$$\|A\|_F = \left( \sum_{i=1}^{m} \sum_{j=1}^{n} a_{ij}^2 \right)^{1/2}.$$

(2.14)

which satisfies

$$\|A\|_F^2 = \text{trace}(A^T A) = \text{trace}(A A^T) = \sum_{j=1}^{n} \sigma_j^2.$$

(2.15)

Other examples are the Schatten p-norms,

$$\|A\|_p = \left( \sum_{j=1}^{n} \sigma_j^p \right)^{1/p}, \quad 1 \leq p < \infty$$

(2.16)

and Ky Fan k-norms,

$$\|A\|_{(k)} = \sum_{j=1}^{k} \sigma_j, \quad k = 1, \cdots, n.$$

(2.17)

The trace norm,

$$\|A\|_{\mathrm{tr}} = \sum_{j=1}^{n} \sigma_j$$

(2.18)

is obtained for k=n and p=1, while the spectral norm

$$\|A\|_{\mathrm{sp}} = \sigma_1 = \max_j \sigma_j$$

(2.19)

corresponds to k=1 and $p = \infty$. The use of Ky Fan k-norms enables us to state the dominance principle in the following way.

Theorem 1 (Ky Fan dominance theorem) The Inequality (2.13) holds for any unitarily invariant norm if and only if

$$\|B\|_{(k)} \le \|C\|_{(k)} \quad \text{for } k = 1, \cdots, n.$$

(2.20)

Another useful tool is the following "rectangular" version of Cauchy interlacing theorem. For a proof of this result see ([24], p. 229) or ([25], p. 1250).

Theorem 2 (A rectangular Cauchy interlace theorem) Let the $m \times n$ matrix A have the singular values (2.2). Let the $\tilde{m} \times \tilde{n}$ matrix $\tilde{A}$ be a submatrix of A which is obtained by deleting m' rows and n' columns of A. That is, and $\tilde{n} \times n' = n$. Define $\tilde{k} \times \min\{\tilde{m}, \tilde{n}\}$ and let

$$\tilde{\sigma}_1 \ge \tilde{\sigma}_2 \ge \cdots \ge \tilde{\sigma}_{\tilde{k}} \ge 0$$

denote the singular values of $\tilde{A}$. Then

$$\sigma_j \geq \tilde{\sigma}_j \geq \sigma_{m'+n'+j} \quad \text{for } j = 1, \cdots, \tilde{k}.$$

(2.21)

To ease the coming discussions we return to square matrices. In the next assertions $W = (w_{ij}) \in \mathbb{R}^{n \times n}$ is an arbitrary real $n \times n$ matrix. Combining the interlace theorem with the dominance theorem leads to the following corollary.

**Theorem 3** Let the $n \times n$ matrix $B_k$ be obtained from W by setting to zero all the entries in the last n-k rows and columns of W. Then the inequality

$$\|B_k\| \leq \|W\|$$

(2.22)

holds for any unitarily invariant norm.

**Theorem 4** Let the $n \times n$ diagonal matrix

$$\text{diag}(W) = \text{diag}\{w_{11}, \ldots, w_{nn}\},$$

be obtained from the diagonal entries of W. Then

$$\|\text{diag}(W)\| \leq \|W\|$$

(2.23)

in any unitarily invariant norm.

Proof. There is no loss of generality in assuming that the diagonal entries of W are ordered such that

$$\left|w_{11}\right| \geq \left|w_{22}\right| \geq \cdots \geq \left|w_{nn}\right|.$$

Let the matrix $B_k$ be defined as in Theorem 3. Then from (2.10) and (2.22) we conclude that

$$\left\|\operatorname{diag}(W)\right\|_{(k)} \leq \left\|B_k\right\|_{(k)} \leq \left\|W\right\|_{(k)} \qquad \text{for } k = 1, \cdots, n,$$

which proves (2.23).

Corollary 5 The diagonal matrix

$$\operatorname{diag}(B_k) = \operatorname{diag}\left\{w_{11}, \cdots, w_{kk}, 0, 0, \cdots, 0\right\}$$

satisfies

$$\left\|\operatorname{diag}(B_k)\right\| \leq \left\|\operatorname{diag}(W)\right\| \leq \left\|W\right\| \qquad (2.24)$$

in any unitarily invariant norm.

Lemma 6 Let X and Y be a pair of real symmetric $n \times n$ matrices that satisfy

$$0 \leq X \leq Y.$$

Then

$$\|X\| \leq \|Y\|$$

in any unitarily invariant norm.

Proof. Using the spectral decomposition of X it is possible to assume that X is a diagonal matrix:

$$X = \operatorname{diag}\{\lambda_1, \cdots, \lambda_n\}.$$

The matrix Y-X is positive semidefinite and, therefore, has non-negative diagonal entries. This observation implies the inequalities

$$y_{jj} \geq \lambda_j \quad \text{for } j = 1, \cdots, n,$$

and

$$\|X\| = \left\|\operatorname{diag}\{\lambda_1, \cdots, \lambda_n\}\right\| \leq \left\|\operatorname{diag}\{y_{11}, \cdots, y_{nn}\}\right\| = \left\|\operatorname{diag}(Y)\right\|,$$

while (2.23) gives

$$\left\|\operatorname{diag}(Y)\right\| \leq \|Y\|.$$

Theorem 7 (The pinching principle) Let the matrix $W \in \mathbb{R}^{n \times n}$ be partitioned in the form

$$\left(\begin{array}{c|c} W_{11} & W_{12} \\ \hline W_{21} & W_{22} \end{array}\right)$$

$$(2.25)$$

where $W_{11} \in \mathbb{R}^{q \times q}$ and $W_{22} \in \mathbb{R}^{(n-q) \times (n-q)}$. Let the $n \times n$ matrix

$$\hat{W} = \left( \begin{array}{c|c} W_{11} & 0 \\ \hline 0 & W_{22} \end{array} \right)$$

(2.26)

denote the "pinched" version of W. Then the inequality

$$\|\hat{W}\| \le \|W\|$$

(2.27)

holds in any unitarily invariant norm.

Proof. Using the SVD of $W_{11}$ we obtain an pair of $q \times q$ orthonormal matrices, $U_{11}$ and $V_{11}$, such that

$U_{11}^T W_{11} V_{11}$ is a diagonal matrix that contains the singular values of $W_{11}$. Similarly there exists a pair of $(n-q) \times (n-q)$ orthonormal matrices, $U_{22}$ and $V_{22}$, such that $U_{22}^T W_{22} V_{22}$ is a diagonal matrix that contains

the singular values of $W_{22}$. The related $n \times n$ matrices

$$U = \left( \begin{array}{c|c} U_{11} & 0 \\ \hline 0 & U_{22} \end{array} \right) \quad \text{and} \quad V = \left( \begin{array}{c|c} V_{11} & 0 \\ \hline 0 & V_{22} \end{array} \right)$$

are orthonormal matrices, and

$$U^T \hat{W} V = \left( \begin{array}{c|c} U_{11}^T W_{11} V_{11} & \\ \hline & U_{22}^T W_{22} V_{22} \end{array} \right)$$

(2.28)

is a diagonal matrix. Moreover, comparing $U^T W V$ with $U^T \hat{W} V$ shows that

$$U^{\mathrm{T}}\hat{W}V = \mathrm{diag}\left(U^{\mathrm{T}}WV\right).$$

(2.29)

Hence a further use of (2.23) gives

$$\left\|\hat{W}\right\| = \left\|U^{\mathrm{T}}\hat{W}V\right\| = \left\|\mathrm{diag}\left(U^{\mathrm{T}}WV\right)\right\| \le \left\|U^{\mathrm{T}}WV\right\| = \left\|W\right\|.$$

Equality (2.28) relates the singular values of $\hat{W}$ with those of the matrices $W_{11}$ and $W_{22}$: Each singular value of $W_{11}$ is a singular value of $\hat{W}$. Similarly, each singular value of $W_{22}$ is a singular value of $\hat{W}$. Conversely, each singular value of $\hat{W}$ is a singular value of $W_{11}$ or a singular value of $W_{22}$. The last observation enables us to sharpen the results in certain cases. This is illustrated in Lemmas 8-11 below, which seem to be new. We will use these lemmas in the proofs of Theorems 18-21.

Lemma 8 (Pinching in Schatten p-norms)

$$\left\|W\right\|_p^p \ge \left\|\hat{W}\right\|_p^p = \left\|W_{11}\right\|_p^p + \left\|W_{22}\right\|_p^p.$$

(2.30)

Lemma 9 (Pinching in the trace norm)

$$\left\|W\right\|_{\mathrm{tr}} \ge \left\|\hat{W}\right\|_{\mathrm{tr}} = \left\|W_{11}\right\|_{\mathrm{tr}} + \left\|W_{22}\right\|_{\mathrm{tr}}.$$

(2.31)

Lemma 10 (Pinching in the spectral norm)

$$\left\|W\right\|_{\mathrm{sp}} \ge \left\|\hat{W}\right\|_{\mathrm{sp}} = \max\left\{\left\|W_{11}\right\|_{\mathrm{sp}}, \left\|W_{22}\right\|_{\mathrm{sp}}\right\}.$$

(2.32)

Lemma 11 (Pinching in Ky Fan k-norms) Let $k_1$ and $k_2$ be a pair of positive integers that satisfy

$$1 \le k_1 \le q \quad \text{and} \quad 1 \le k_2 \le n - q.$$

Then

$$\left\|W\right\|_{(k_1+k_2)} \geq \left\|\hat{W}\right\|_{(k_1+k_2)} \geq \left\|W_{11}\right\|_{(k_1)} + \left\|W_{22}\right\|_{(k_2)}.$$

(2.33)

Proof. The sum $\left\|W_{11}\right\|_{(k_1)} + \left\|W_{22}\right\|_{(k_2)}$ is formed from $k_1+k_2$ singular values of $\hat{W}$, while the sum defined by $\left\|\hat{W}\right\|_{(k_1+k_2)}$ is composed from the $k_1+k_2$ largest singular values of $\hat{W}$.

The next tools consider the problem of approximating one matrix by another matrix of lower rank. Let $A \in \mathbb{R}^{m \times n}$ by a given matrix with SVD that satisfies (2.1)-(2.8). Let $1 \leq k \leq n$ be a given integer, and let

$$\mathbb{B}_k = \left\{ B \mid B \in \mathbb{R}^{m \times n} \ \text{and} \ \text{rank}(B) \leq k \right\}$$

denote the related set of low-rank matrices. Then here we seek a matrix $B \in \mathbb{B}$ that is nearest to A in a certain matrix norm. The difficulty stems from the fact that $\mathbb{B}_k$ is not a convex set. Let $T_k$ denote a rank-k truncated SVD of A as defined in (2.8). Then the Eckart-Young theorem [26] says that $T_k$ solves this problem in the Frobenius norm. The extension of this result to any unitarily invariant norm is due to Mirsky [27]. (Recall that $T_k$ is not always unique. In such cases the nearest matrix is not unique.) A detailed statement of these assertions is given below. For recent discussions and proofs see [25].

Theorem 12 (Eckart-Young) The inequality

$$\left\|A - B\right\|_F^2 \geq \sum_{j=k+1}^{n} \sigma_j^2$$

holds for any matrix $B \in \mathbb{B}_k$. Moreover, the matrix $T_k$ solves the problem

$$\begin{aligned}
\text{minimize} \quad & F(B) = \|A - B\|_F^2 \\
\text{subject to} \quad & B \in \mathbb{B}_k,
\end{aligned}$$

giving the optimal value of

$$\|A - T_k\|_F^2 = \left\| \sum_{j=k+1}^{n} \sigma_j \mathbf{u}_j \mathbf{v}_j^T \right\|_F^2 = \sum_{j=k+1}^{n} \sigma_j^2.$$

**Theorem 13 (Mirsky)** Let $\|\cdot\|$ be any unitarily invariant norm on $\mathbb{R}^{m \times n}$. Then the inequality

$$\|A - B\| \ge \|A - T_k\|$$

holds for any matrix $B \in \mathbb{B}_k$. In other words, the matrix $T_k$ solves the problem

$$\begin{aligned}
\text{minimize} \quad & \mu(B) = \|A - B\| \\
\text{subject to} \quad & B \in \mathbb{B}_k.
\end{aligned}$$

## POSITIVE APPROXIMANTS OF SYMMETRIC MATRICES

In this section we consider the solution of problem (1.3). Since $\|\cdot\|$ is a unitarily invariant norm, the spectral decomposition (1.2) enables us to convert (1.3) into the simpler form

$$\begin{aligned}
\text{minimize} \quad & F(S) = \|\Lambda - S\| \\
\text{subject to} \quad & S \in \mathbb{S}_n^+,
\end{aligned} \tag{3.1}$$

whose solution provides a solution of (1.3).

Theorem 14 let the matrix $\Lambda_q$ be defined as in (1.5). Then $\Lambda_q$ solves (3.1) in any unitarily invariant norm.

Proof. Let the diagonal matrix $D_q$ be defined by the equality

$$\Lambda + D_q = \Lambda_q .$$

That is

$$D_q = \operatorname{diag}\left\{0,\cdots,0,\left|\lambda_{q+1}\right|,\cdots,\left|\lambda_n\right|\right\}.$$

$$(3.2)$$

Let $S = (s_{ij})$ be some matrix in $\mathbb{S}_n^+$ and let the matrix $W = (w_{ij}) \in \mathbb{R}^{n \times n}$ be defined by the equality

$$\Lambda + W = S.$$

$$(3.3)$$

Then the proof is concluded by showing that

$$\|W\| \ge \|D_q\|.$$

$$(3.4)$$

Let the diagonal matrix

$$W_q = \operatorname{diag}\left\{0,\cdots,0,w_{q+1,q+1},\cdots,w_{nn}\right\}$$

$$(3.5)$$

be obtained from the last n-q diagonal entries of W. Then Corollary 5 implies that

$$\|W\| \ge \|W_q\|.$$

(3.6)

On the other hand, since $S \ge 0$, the diagonal entries of S are non-negative, which implies the inequalities

$$w_{jj} \ge |\lambda_j| \quad \text{for } j = q+1, \cdots, n,$$

(3.7)

and

$$\|W_q\| \ge \|D_q\|.$$

(3.8)

Now combining (3.6) and (3.8) gives (3.4)

Theorem 14 is not new, e.g. ([8], p. 277) and [9]. However, the current proof is simple and short. In the next sections we extend these arguments to derive low-rank approximants.

## LOW-RANK POSITIVE APPROXIMANTS IN THE FROBENIUS NORM

In this section we consider the solution of problem (1.1) in the Frobenius norm. As before, the spectral decom- position (1.2) can be used to "diagonalize" the problem and the actual problem to solve has the form

$$\text{minimize} \quad F(S) = \|\Lambda - S\|_F^2$$

$$\text{subject to} \quad S \in \mathbb{S}_{n,\ell}^+.$$

(4.1)

Theorem 15 Let the matrix $\Lambda_l$ be defined as in (1.6). Then this matrix solves (4.1)

Proof. Let the diagonal matrix $D_\ell$ be defined by the equality

$$\Lambda + D_\ell = \Lambda_\ell.$$

That is,

$$D_\ell = \operatorname{diag}\left\{0,\cdots,0,-\lambda_{\ell+1},\cdots,-\lambda_q,\left|\lambda_{q+1}\right|,\cdots,\left|\lambda_n\right|\right\},$$

(4.2)

and

$$\|D_\ell\|_F^2 = \sum_{j=\ell+1}^{q}\lambda_j^2 + \sum_{j=q+1}^{n}\lambda_j^2.$$

(4.3)

Let $S = (s_{ij})$ be some matrix in $\mathbb{S}_{n,l}^{+}$ and let the matrix $W = (w_{ij}) \in \mathbb{R}^{n \times n}$ be defined by the equality

$$\Lambda + W = S.$$

(4.4)

Then the proof is concluded by showing that

$$\|W\|_F^2 \geq \|D_\ell\|_F^2.$$

(4.5)

This aim is achieved by considering a partition of W and S in the form

$$W = \left(\begin{array}{c|c} W_{11} & W_{12} \\ \hline W_{21} & W_{22} \end{array}\right) \quad \text{and} \quad S = \left(\begin{array}{c|c} S_{11} & S_{12} \\ \hline S_{21} & S_{22} \end{array}\right)$$

(4.6)

where $W_{11}$ and $S_{11}$ are $q \times q$ matrices, while $W_{22}$ and $S_{22}$ are $(n-q) \times (n-q)$ matrices. Then, clearly,

$$\|W\|_F^2 \geq \|W_{11}\|_F^2 + \|W_{22}\|_F^2 \, . \tag{4.7}$$

Also, as before, since S is a positive semidefinite matrix it has non-negative diagonal entries, which implies the inequalities

$$w_{jj} \geq |\lambda_j| \quad \text{for } j = q+1, \cdots, n \tag{4.8}$$

and

$$\|W_{22}\|_F^2 \geq \sum_{j=q+1}^{n} w_{jj}^2 \geq \sum_{j=q+1}^{n} \lambda_j^2 . \tag{4.9}$$

It is left, therefore, to show that

$$\|W_{11}\|_F^2 \geq \sum_{j=\ell+1}^{q} \lambda_j^2 . \tag{4.10}$$

Observe that the matrices $W_{11}$ and $S_{11}$ are related by the equality

$$\Lambda_{11} + W_{11} = S_{11} \tag{4.11}$$

where

$$\Lambda_{11} = \text{diag}\{\lambda_1, \cdots, \lambda_q\} \in \mathbb{R}^{q \times q} \tag{4.12}$$

and

$$\lambda_1 \geq \cdots \geq \lambda_q > 0. \tag{4.13}$$

Moreover, since $S_{11}$ is a principal submatrix of S,

$$\text{rank}\left(S_{11}\right) \le \text{rank}\left(S\right) \le \ell. \tag{4.14}$$

Hence from the Eckart-Young theorem we obtain that

$$\left\|W_{11}\right\|_F^2 = \left\|\Lambda_{11} - S_{11}\right\|_F^2 \ge \sum_{j=\ell+1}^q \lambda_j^2. \tag{4.15}$$

Corollary 16 Let X be a given real symmetric $n \times n$ matrix with the spectral decomposition (1.2). Then the matrix

$$X_\ell = Q\Lambda_\ell Q^T \tag{4.16}$$

solves the problem

$$\begin{aligned}
&\text{minimize} && F\left(S\right) = \left\|X - S\right\|_F^2 \\
&\text{subject to} && S \in \mathbb{S}_{n,\ell}^+.
\end{aligned} \tag{4.17}$$

Corollary 17 Let $A \in \mathbb{R}^{n \times n}$ be a given matrix, let the matrix $X = (A + A^T)/2$ have the spectral decomposition (1.2), and let the matrix $X_\ell$ be defined in (4.16). Then $X_\ell$ solves the problem

$$\begin{aligned}
&\text{minimize} && G\left(S\right) = \left\|A - S\right\|_F^2 \\
&\text{subject to} && S \in \mathbb{S}_{n,\ell}^+.
\end{aligned} \tag{4.18}$$

## LOW-RANK POSITIVE APPROXIMANTS IN THE SCHATTEN P-NORM

Let the diagonal matrix $\Lambda$ be obtained from the spectral decomposition (1.2). In this section we consider the problem minimize $\quad F(S) = \|\Lambda - S\|_p^p$

subject to $\quad S \in \mathbb{S}_{n,\ell}^+.$ (5.1)

Theorem 18 Let the matrix $\Lambda_l$ be defined in (1.6). Then this matrix solves (5.1)

Proof. Let the matrices $D_l, W$, and S be defined as in the proof of Theorem 15. Then here it is necessary to prove that

$$\|W\|_p^p \geq \|D_\ell\|_p^p,$$ (5.2)

where

$$\|D_\ell\|_p^p = \sum_{j=\ell+1}^{q} \lambda_j^p + \sum_{j=q+1}^{n} |\lambda_j|^p.$$ (5.3)

Let W and S be partitioned as in (4.6). Then from Lemma 8 we have

$$\|W\|_p^p \geq \|W_{11}\|_p^p + \|W_{22}\|_p^p.$$ (5.4)

Now Theorem 4 and (4.8) imply

$$\|W_{22}\|_p^p \geq \|\mathrm{diag}(W_{22})\|_p^p = \sum_{j=q+1}^{n} |w_{jj}|^p \geq \sum_{j=q+1}^{n} |\lambda_j|^p,$$ (5.5)

while applying Mirsky theorem on (4.11)-(4.14) gives

$$\|W_{11}\|_p^p = \|\Lambda_{11} - S_{11}\|_p^p \geq \sum_{j=\ell+1}^{q} \lambda_j^p.$$ (5.6)

Finally substituting (5.5) and (5.6) into (5.4) gives (5.2).

## LOW-RANK POSITIVE APPROXIMANTS IN THE TRACE NORM

Using the former notations, here we consider the problem

minimize $\quad F(S) = \|\Lambda - S\|_{\mathrm{tr}}$

subject to $\quad S \in \mathbb{S}_{n,\ell}^{+}.$ (6.1)

Theorem 19 The matrix $\Lambda_\ell$ solves (6.1).

Proof. It is needed to show that

$$\|W\|_{\mathrm{tr}} \geq \|D_\ell\|_{\mathrm{tr}}$$ (6.2)

where

$$\|D_\ell\|_{\mathrm{tr}} = \sum_{j=\ell+1}^{q} |\lambda_j| + \sum_{j=q+1}^{n} |\lambda_j|.$$ (6.3)

The use of Lemma 9 yields

$$\|W\|_{\mathrm{tr}} \geq \|W_{11}\|_{\mathrm{tr}} + \|W_{22}\|_{\mathrm{tr}}.$$ (6.4)

Here Theorem 4 and (4.8) imply the inequalities

$$\|W_{22}\|_{\mathrm{tr}} \geq \|\mathrm{diag}(W_{22})\|_{\mathrm{tr}} = \sum_{j=q+1}^{n} |w_{jj}| \geq \sum_{j=q+1}^{n} |\lambda_j|,$$ (6.5)

and Mirsky theorem gives

$$\left\|W_{11}\right\|_{\mathrm{tr}} = \left\|\Lambda_{11} - S_{11}\right\|_{\mathrm{tr}} \geq \sum_{j=\ell+1}^{q} \left|\lambda_n\right|,$$

$$(6.6)$$

which completes the proof.

## LOW-RANK POSITIVE APPROXIMANTS IN THE SPECTRAL NORM

In this case we consider the problem

$$\text{minimize} \quad F(S) = \left\|\Lambda - S\right\|_{\mathrm{sp}}$$

$$\text{subject to} \quad S \in \mathbb{S}_{n,\ell}^{+}.$$

$$(7.1)$$

Theorem 20 The matrix $\Lambda_l$ solves (7.1).

Proof. Following the former notations and arguments, here it is needed to show that

$$\left\|W\right\|_{\mathrm{sp}} \geq \left\|D_\ell\right\|_{\mathrm{sp}}.$$

Define

$$\alpha = \max_{j=\ell+1,\cdots,q} \left|\lambda_j\right| \quad \text{and} \quad \beta = \max_{j=q+1,\cdots,n} \left|\lambda_j\right|.$$

Then, clearly,

$$\left\|D_\ell\right\|_{\mathrm{sp}} = \max_{j=\ell+1,\cdots,n} \left|\lambda_j\right| = \max\{\alpha,\beta\}.$$

Using Lemma 10 we see that

$$\|W\|_{sp} \geq \max\left\{\|W_{11}\|_{sp}, \|W_{22}\|_{sp}\right\}.$$

Now Theorem 4 and (4.8) imply

$$\|W_{22}\|_{sp} \geq \|\mathrm{diag}\,(W_{22})\|_{sp} = \max_{j=q+1,\cdots,n}|w_{jj}| \geq \max_{j=q+1,\cdots,n}|\lambda_j| = \beta,$$

while Mirsky theorem gives

$$\|W_{11}\|_{sp} = \|\Lambda - S_{11}\|_{sp} \geq \max_{j=\ell+1,\cdots,q}|\lambda_j| = \alpha.$$

## UNITARILY INVARIANT NORMS

Let the diagonal matrices $\Lambda$ and $\Lambda_I$ be defined as in Section 1, and let $\|\cdot\|$ denote any unitarily invariant norm on $\mathbb{R}^{n\times n}$. Below we will show that $\Lambda_I$ solves the problem

$$\begin{aligned} \text{minimize} \quad & F(S) = \|\Lambda - S\| \\ \text{subject to} \quad & S \in \mathbb{S}^+_{n,\ell}. \end{aligned} \tag{8.1}$$

The derivation of this result is based on the following assertion, which considers Ky Fan k-norms.

Theorem 21 The matrix $\Lambda_I$ solves the problem

minimize $\quad F(S) = \|\Lambda - S\|_{(k)}$

subject to $\quad S \in \mathbb{S}_{n,\ell}^{+}$

$$(8.2)$$

For $k = 1, \cdots, n$.

Proof. We have already proved that $\Lambda_I$ solves (8.2) for the spectral norm (k=1) and the trace norm (k=n). Hence it is left to consider the case when $2 \leq k \leq n-1$. As before, the diagonal matrix $D_I$ is defined in (4.2), and the matrices S and W satisfy (4.4) as well as the partition (4.6). With these notations at hand it is needed to show that

$$\|W\|_{(k)} \geq \|D_\ell\|_{(k)},$$

$$(8.3)$$

Let $D_I$ be partitioned in a similar way:

$$D_\ell = \left( \begin{array}{c|c} D_{11} & 0 \\ \hline 0 & D_{22} \end{array} \right),$$

$$(8.4)$$

where

$$D_{11} = \text{diag}\left\{0, \cdots, 0, -\lambda_{\ell+1}, \cdots, -\lambda_q\right\} \in \mathbb{R}^{q \times q}$$

$$(8.5)$$

and

$$D_{22} = \text{diag}\left\{|\lambda_{q+1}|, \cdots, |\lambda_n|\right\} \in \mathbb{R}^{(n-q) \times (n-q)}.$$

$$(8.6)$$

Then there are three different cases to consider.

The first case occurs when

$$\left\| D_\ell \right\|_{(k)} = \left\| D_{11} \right\|_{(k)}.$$

(8.7)

Here Theorem 3 implies the inequalities

$$\left\| W \right\|_{(k)} \geq \left\| W_{11} \right\|_{(k)},$$

(8.8)

while from (4.11)-(4.14) and Mirsky theorem we obtain

$$\left\| W_{11} \right\|_{(k)} \geq \left\| D_{11} \right\|_{(k)} = \left\| D_\ell \right\|_{(k)},$$

(8.9)

which proves (8.3).

The second case occurs when

$$\left\| D_\ell \right\|_{(k)} = \left\| D_{22} \right\|_{(k)}.$$

(8.10)

Here Theorem 3 implies

$$\left\| W \right\|_{(k)} \geq \left\| W_{22} \right\|_{(k)},$$

(8.11)

while Theorem 4 and the inequalities (4.8) give

$$\left\| W_{22} \right\|_{(k)} \geq \left\| \mathrm{diag}\left( W_{22} \right) \right\|_{(k)} \geq \left\| D_{22} \right\|_{(k)} = \left\| D_\ell \right\|_{(k)},$$

(8.12)

which proves (8.3).

The third case occurs when neither (8.7) nor (8.10) hold. In this case there exist two positive integers, $k_1$ and $k_2$, such that

$$k_1 + k_2 = k \tag{8.13}$$

and

$$\left\| D_\ell \right\|_{(k)} = \left\| D_{11} \right\|_{(k_1)} + \left\| D_{22} \right\|_{(k_2)}. \tag{8.14}$$

Now Lemma 11 shows that

$$\left\| W \right\|_{(k)} \geq \left\| W_{11} \right\|_{(k_1)} + \left\| W_{22} \right\|_{(k_2)}. \tag{8.15}$$

A further use of (4.11)-(4.14) and Mirsky theorem give

$$\left\| W_{11} \right\|_{(k_1)} \geq \left\| D_{11} \right\|_{(k_1)}, \tag{8.16}$$

and from Theorem 4 and (4.8) we obtain

$$\left\| W_{22} \right\|_{(k_2)} \geq \left\| \mathrm{diag}\left( W_{22} \right) \right\|_{(k_2)} \geq \left\| D_{22} \right\|_{(k_2)}. \tag{8.17}$$

Hence by substituting (8.16) and (8.17) into (8.15) we get (8.3).

The fact that (8.3) holds for $k = 1, \cdots, n$ means that the inequality

$$\left\| W \right\| \geq \left\| D_\ell \right\| \tag{8.18}$$

holds for any unitarily invariant norm. This observation is a direct consequence of Ky Fan dominance theorem. The last inequality proves our final results.

Theorem 22 The matrix $\Lambda_l$ solves (8.1) in any unitarily invariant norm.

Theorem 23 Using the notations of Section 1, the matrix

$$X_\ell = Q\Lambda_\ell Q^\mathsf{T}$$

solves (1.1) in any unitarily invariant norm.

## CONCLUDING REMARKS

In view of Theorem 14 and Mirsky theorem, the observation that $\Lambda_l$ solves (8.1) is not surprising. However, as we have seen, the proof of this assertion is not straightforward. A key argument in the proof is the inequality (8.15), which is based on Lemma 11.

Once Theorem 22 is proved, it is possible to use this result to derive Theorems 15-18. Yet the direct proofs that we give clearly illustrate why these theorems work. In fact, the proof of Theorem 15 paves the way for the other proofs. Moreover, as Corollary 17 shows, when using the Frobenius norm we get stronger results: In this case we are able to compute a low-rank positive approximant of any matrix $A \in \mathbb{R}^{n \times n}$.

## REFERENCES

1. Dax, A. (2014) Imputing Missing Entries of a Data Matrix: A Review. Techical Report, Hydrological Service of Israel.

2. Trosset, M.W. (2000) Distance Matrix Completion by Numerical Optimization. Computational Optimization and Applications, 17, 11-22. http://dx.doi.org/10.1023/A:1008722907820

3. Higham, N.J. (1988) Computing a Nearest Symmetric Positive Semidefinite Matrix. Linear Algebra and Its Applications, 103, 103-118. http://dx.doi.org/10.1016/0024-3795(88)90223-6

4. Halmos, P.R. (1972) Positive Approximants of Operators. Indiana University Mathematics Journal, 21, 951-960. http://dx.doi.org/10.1512/iumj.1972.21.21076

5. Rogers, D.D. and Ward, J.D. (1981) $C_p$-Minimal Positive Approximants. Acta Scientiarum Mathematicarum, 43, 109-115.

6. Ando, T. (1985) Approximation in Trace Norm by Positive Semidefinite Matrices. Linear Algebra and Its Applications, 71, 15-21. http://dx.doi.org/10.1016/0024-3795(85)90230-7

7. Ando, T., Sekiguchi, T. and Suzuki, T. (1973) Approximation by Positive Operators. Mathematische Zeitschrift, 131, 273-282. http://dx.doi.org/10.1007/BF01174903

8. Bhatia, R. (1997) Matrix Analysis. Springer, New York. http://dx.doi.org/10.1007/978-1-4612-0653-8

9. Bhatia, R. and Kittaneh, F. (1992) Approximation by Positive Operators. Linear Algebra and Its Applications, 161, 1-9. http://dx.doi.org/10.1016/0024-3795(92)90001-Q

10. Bouldin, R. (1973) Positive Approximants. Transactions of the American Mathematical Society, 177, 391-403. http://dx.doi.org/10.1090/S0002-9947-1973-0317082-6

11. Sekiguchi, T. (1976) Positive Approximants of Normal Operators. Hokkaido Mathematical Journal, 5, 270-279. http://dx.doi.org/10.14492/hokmj/1381758677

12. Bouldin, R. (1980) Best Approximation of a Normal Operator in the Schatten p-Norm. Proceedings of the American Mathematical Society, 80, 277-282.

13. Bouldin, R. (1987) Best Approximation of a Normal Operator in the Trace Norm. Acta Scientiarum Mathematicarum, 51, 467-474.

14. Bouldin, R., (1988) Best Approximation of a Nonnormal Operator in the Trace Norm. Journal of Mathematical Analysis and Applications, 132, 102-113. http://dx.doi.org/10.1016/0022-247X(88)90046-7

15. Laszkiewicz, B. and Zietak, K. (2008) Approximation by Matrices with Restricted Spectra. Linear Algebra and Its Applications, 428, 1031-1040. http://dx.doi.org/10.1016/j.laa.2007.09.009

16. Phillips, J. (1977) Nearest Normal Approximation for Certain Normal Operators. Proceedings of the American Mathematical Society, 67, 236-240. http://dx.doi.org/10.1090/S0002-9939-1977-0458212-4

17. Ruhe, A. (1987) Closest Normal Matrix Finally Found! BIT Numerical Mathematics, 27, 585-598. http://dx.doi.org/10.1007/BF01937278

18. Zietak, K. (1997) Strict Spectral Approximation of a Matrix and Some Related Problems. Applicationes Mathematicae, 24, 267-280.

19. Higham, N.J. (1989) Matrix Nearness Problems and Applications. In: Gover, M.J.C. and Barnett, S., Eds., Applications of Matrix Theory, Oxford University Press, Oxford, 1-27.

20. Fan, K. (1951) Maximum Properties and Inequalities for the Eigenvalues of Completely Continuous Operators. Proceedings of the National Academy of Sciences of the United States of America, 37, 760-766. http://dx.doi.org/10.1073/pnas.37.11.760

21. Fan, K. and Hoffman, A.J. (1955) Some Metric Inequalities in the Space of Matrices. Proceedings of the American Mathematical Society, 6, 111-116.http://dx.doi.org/10.1090/S0002-9939-1955-0067841-7

22. Horn, R.A. and Johnson, C.R. (1991) Topics in Matrix Analysis. Cambridge University Press, Cambridge. http://dx.doi.org/10.1017/CBO9780511840371

23. Marshall, A.W., Olkin, I. and Arnold, B.C. (2011) Inequalities: Theory of Majorization and Its Applications. Springer Series in Statistics, 2nd Edition, Springer, New York.

24. Zhang, F. (1999) Matrix Theory: Basic Results and Techniques. Springer-Verlag, New York. http://dx.doi.org/10.1007/978-1-4757-5797-2

25. Dax, A. (2010) On Extremum Properties of Orthogonal Quotient Matrices. Linear Algebra and Its Applications, 432, 1234-1257. http://dx.doi.org/10.1016/j.laa.2009.10.034

26. Eckart, C. and Young, G. (1936) The Approximation of One Matrix by Another of Lower Rank. Psychometrika, 1, 211-218. http://dx.doi.org/10.1007/BF02288367

27. Mirsky, L. (1960) Symmetric Gauge Functions and Unitarily Invariant Norms. Quarterly Journal of Mathematics, 11, 50-59. http://dx.doi.org/10.1093/qmath/11.1.50

## CITATION

Dax, A. (2014) Low-Rank Positive Approximants of Symmetric Matrices. Advances in Linear Algebra & Matrix Theory, 4, 172-185. doi: 10.4236/alamt.2014.43015.

# On the Initial Sub algebra of a Graded Lie Algebra

## Thomas B. Gregory
Department of Mathematics, the Ohio State University at Mansfield, Mansfield, Ohio, USA

## ABSTRACT

We show that each irreducible, transitive finite-dimensional graded Lie algebra over a field of prime characteristic p contains an initial sub algebra in which the $p^{th}$ power of the adjoin transformation associated with any element in the lowest gradation space is zero.

## INTRODUCTION

In the classification of the simple finite-dimensional Lie algebras over fields of prime characteristic, irreducible transitive finite dimensional graded Lie algebras play a fundamental role [1]. The simple finite dimensional Lie algebras over algebraically closed fields of characteristic greater than three have been classified [2]. Work is being done in characteristic three [3] - [7]. It is well known that in Lie algebras of Cartan type, there is a (not necessarily proper) subalgebra, the "initial piece," which contains the sum of the negative gradations spaces of the Lie algebra, and in which the $P^{th}$ power of the adjoint representation associated with any element of the lowest gradation space is zero. In this paper, we prove that any irreducible, transitive finite-dimensional graded Lie algebra contains such an initial subalgebra. Indeed, we prove the following theorem.

## MAIN THEOREM

Let

$$G = G_{-q} \oplus \cdots \oplus G_{-1} \oplus G_0 \oplus G_1 \oplus \cdots \oplus G_r, \quad r \geq 1, \quad q \geq 1$$

be an irreducible, transitive, finite-dimensional graded Lie algebra over a field of characteristic $P$ such that $M(G) = 0$ [8]. Then $G$ contains an irreducible, transitive depth $_-q$ graded sub algebra

$$R = \prod_{0 \leq i \leq l} (\mathrm{ad}\, v_i)^{(p-1)p^{j}} G + G_0,$$

Where $v_i \in G_{-}q$, and where I is a non-negative whole number. We have $G_{-}q \subseteq R, G_{-1} \subseteq R$, and $(\mathrm{ad}\, v)^P R = 0$ for all $v \in G_{-q}$.

If $q \geq r$, then the conclusion of the theorem obviously holds. In what follows, therefore, we will assume that $r > q$.

### Intermediate Results

To prove the Main Theorem, we will make use of the following series of lemmas, in which we assume the hypotheses and notation of the Main Theorem. We note that by, for example, [9] (Lemma 6), G is transitive in its negative part. (Note that the lemmas we quote from [9] are valid for all prime characteristics.) As usual, we assume throughout that M (G) = 0 [8].

***Lemma 1.*** If M is an abelian $G_0$-sub module of G, then for any $m \in M$, $(\mathrm{ad}\, m)^{p^{j}}$ is a $G_0$-endomorphism of G for all $j > 0$.

***Proof.*** For any $g \in G_0, m \in M$, and $x \in G$, we have

$$[m, [m, g]] \subseteq [m, M] = 0$$

So that modulo P

**Lemma 2.** If $m \in G$ is such that $(\text{ad } m)^{p^j} G_o = o$ for some $j > 0.$, then $(\text{ad } m)^{p^j}$ is a $G_o$-endomorphism of $G$.

**Proof.** As in the proof of Lemma 1 above, we have, for any $g \in G_0$ and any $x \in G$, that modulo P

$$(\text{ad } m)^{p^j} [g,x] = \sum_{0 \le k \le p^j} \binom{p^j}{k} \left[(\text{ad } m)^{p^j-k} g.(\text{ad } m)^k x\right] = \left[(\text{ad } m)^{p^j} g,x\right] + \left[g.(\text{ad } m)^{p^j} x\right]$$

$$= [0,x] + \left[g.(\text{ad } m)^{p^j} x\right] = \left[g.(\text{ad } m)^{p^j} x\right].$$

**Lemma 3.** If $v \in G_{-q}$, and $j > 0.$ is maximal such that $(\text{ad } m)^{p^j} \ne o$, then $(\text{ad } m)^{p^j} G_{p^j q}$ is a Lie sub algebra.

**Proof.** Let $x_1$ and $x_2$ be any elements of $G_{p^j q}$, Then for any $v \in G_{-q}$,

$$\left[(\text{ad } v)^{p^j} x_1, (\text{ad } v)^{p^j} x_2\right] = (\text{ad } v)^{p^j} \left[(\text{ad } v)^{p^j} x_1, x_2\right]$$

Since, as we have seen in the proofs of the previous lemmas, $(\text{ad } v)^{p^j}$ is a derivation, and

$\left((\text{ad } v)^{p^j}\right)^2 x_1 \in G_{-p^j q} = 0$. In addition, since $(\text{ad } v)^{p^j} x_1 \in G_0$ we have $\left[(\text{ad } v)^{p^j} x_1, x_2\right] \in G_{p^j q}$. Hence, $(\text{ad } v)^{p^j} \left[(\text{ad } v)^{p^j} x_1, x_2\right] \in G_{p^j q}$, which, as it is obviously closed under addition, is seen to be a Lie sub algebra, as required;

**Lemma 4.** Let I be the minimal (graded) ideal of G [8]. If $v \in G_{-q}$ is such that $(\text{ad } v)^{p^j} I_k = 0$ for some integers j and k, with $j \ge 0$ and $(p^j -1)q \le m \le r$, then $(\text{ad } v)^{p^j} I_m = 0$ for all m, $-q \le m \le r$, i.e., $(\text{ad } v)^{p^j} I = 0$

**Proof.** Suppose $(ad\ v)^{p^j} I_k = 0$. Then for all m $k \leq m \leq r$, we have (since for all m, $-q \leq m \leq r$, we have $_m = [G_{-1}, I_{m+1}]$)

$$\left(ad\ G_{-1}\right)^{m-k} \left(ad\ v\right)^{p^j} I_m = \left(ad\ v\right)^{p^j} \left(ad\ G_{-1}\right)^{m-k} I_m = \left(ad\ v\right)^{p^j} I_k = 0$$

So $(ad\ v)^{p^j} I_m = 0$ by transitivity. If $m < k$, then

**Lemma 5.** If $(ad\ v)^{p^j} G_k = 0$ for some k such that $p^j q - q \leq m \leq r$ and for some $j > 0$, then $(ad\ v)^{p^j} G = 0$.

**Proof.** We will show that $(ad\ v)^{p^j} G_m = 0$ for all m $p^j q - q \leq m \leq r$. (If $m < p^j q - q$, then

$$(ad\ v)^{p^j} G_m \subseteq 0 \sum\nolimits_{n < -q} G_n = 0)$$ First of all, suppose that $p^j q - q < m \leq r$. Then, since

$(ad\ v)^{p^j} I_k \subseteq (ad\ v)^{p^j} G_k = 0$, we have, by Lemma 4 that $(ad\ v)^{p^j} I = 0$. Consequently, we have

$$0 = (ad\ v)^{pj} I_{m-1} = (ad\ v)^{pj} \left[G_{-1,} (ad\ v)^{pj} G_m\right]$$

So $(ad\ v)^{p^j} G_m = 0$ by the transitivity of G, if $m < p^j q$, or [9] (Lemma 6) otherwise. Finally, if $m < p^j q - q$

And $(ad\ v)^{p^j} G_m \neq 0$, then by Lemma 1 (or Lemma 2), $(ad\ v)^{p^j} G_m$ is a non-zero $G_0$-sub module of $G_{-q}$. But by, for example, [9] (Lemma 9), $G_{-q}$ is irreducible as a $G_0$ module; therefore, $(ad\ v)^{p^j} G_m = G_{-q}$, and we have

$\left[G_{-q}, 1\right] = \left[(ad\,v)^{p^j} G_m, 1\right] = (ad\,v)^{p^j} \left[G_m, 1\right] \subseteq (ad\,v)^{p^j} 1 = 0$ (by Lemma 4, as we noted earlier in the proof). But then, since $G_{-1} \subset 1$, we would have $0 = \left[G_{-q}, l_1\right] \supseteq \left[G_{-q}, [G_{-1}, G_{-2}]\right] = \left[G_{-1}, [G_{-q}, G_2]\right]$ so $\left[G_{-q}, G_2\right] = 0$ by, for example, [9] (Lemma 6), to contradict, for example, [9] (Lemma 8). Thus, we must have $(ad\,v)^{p^j} G_m = 0$ in this case, also, so $(ad\,v)^{p^j} G = 0$ as required.

**Lemma 6.** If $(ad\,v)^{p^j} G \neq 0$ for some $v \in G_{-q}$ and $j > 0$, then both $G_{p^j q}$ and $(ad\,v)^{p^j} G_{p^j q}$ are non-zero, and $G_{-q} \subseteq (ad\,v)^{p^j} G$.

**Proof.** If $(ad\,v)^{p^j} G \neq 0$, then $r \geq p^j q - q$, since otherwise we would have $(ad\,v)^{p^j} G \subseteq \sum_{n < -q} G_n = 0$, contrary to hypothesis. By Lemma 5, $(ad\,v)^{p^j} G_{p^j q - q}$ is not zero, and by Lemma 1 (or Lemma 2), $(ad\,v)^{p^j} G_{p^j q - q}$ is a $G_0$-submodule of $G_{-q}$; hence, by, for example, [9] (Lemma 9),

$$(ad\,v)^{p^j} G_{p^j q - q} = G_{-q}$$

Since $(ad\,v)^{p^j}$ is a derivation of $G$ which annihilates $G_q$, we have, by, for example, [9] (Lemma 8) that $0 \neq \left[G_{-q}, G_q\right] = \left[(ad\,v)^{p^j} G_{p^j q - q}, G_q\right] = (ad\,v)^{p^j} \left[G_{p^j q - q}, G_q\right] \subseteq (ad\,v)^{p^j} G_{p^j q}$.

Thus, both $G_{p^j q}$ and $(ad\,v)^{p^j} G_{p^j q}$ are non-zero, and Lemma 6 is proved.

**Lemma 7.** Let $v$ be a non-zero element of $G_{-q}$. If $j > 0$ is maximal such that $(ad\,v)^{p^j} G \neq 0$, then $(ad\,v)^{(p-1)p^j} G \neq 0$.

**Proof.** Suppose $(\text{ad } v)^{(p-1)p^j}G = 0$. Then for any $x \in G_{p^jq}$, which is non-zero by Lemma 6, we have that

$$0 = \left((\text{ad } v)^{p^j}\right)^{p-1}(\text{ad } x)^{p-1}G_{-1} = (p-1)!\left(\text{ad }(\text{ad } v)^{p^j}x\right)^{p-1}G_{-1}.$$

Thus $\left(\text{ad}(\text{ad } v)^{p^j}x\right)^{p-1}G_{-1} = 0$, so $\text{ad}_{G_{-1}}(\text{ad } v)^{p^j}G_{p^jq}$ is a nil set of endomorphisms of $G_{-1}$. By Lemma 3, this nil set of endomorphism's is weakly closed, so by Jacobson's theorem on nil weakly closed sets

[10], $\text{ad}_{G_{-1}}(\text{ad } v)^{p^j}G_{p^jq}$ acts nil potently on $G_{-1}$ and therefore annihilates

some non-zero element of $G_{-1}$ By Lemma 1 (or Lemma 2), $(\text{ad } v)^{p^j}G_{p^jq}$ is a $G_0$-sub module of $G_0$ (i.e., an ideal of $G_0$). Hence, the annihi-

lator of $(\text{ad } v)^{p^j}G_{p^jq}$ in $G_{-1}$ must be a $G_0$-sub module of $G_{-1}$. By the assumed irreducibility of G, $G_{-1}$ is irreducible as a $G_0$-module. Con-

sequently, $(0 \neq)\text{Anm}_{G_{-1}}(\text{ad } v)^{p^j}G_{p^jq} = G_{-1}$ i.e., $\left[G_{-1}(\text{ad } v)^{p^j}G_{p^jq}\right] = 0$, But

then, we have by transitivity that $(\text{ad } v)^{p^j}G_{p^jq} = 0$, so that, by Lemma 6

again $(\text{ad } v)^{p^j}G = 0$ contrary to the choice of j. Thus, $(\text{ad } v)^{(p-1)p^j}G$ must be non-zero, as asserted.

**Lemma 8.** Let v be a non-zero element of $G_{-q}$, and let $j \geq 0$ be maxi-

mal such that $(\text{ad } v)^{p^j}G \neq 0$. Then $(\text{ad } v)^{(p-1)p^j}G$ is a Lie algebra, and we

have that both $G_{-q} \subseteq (\text{ad } v)^{(p-1)p^j}G$ and $G_{-1} \subseteq (\text{ad } v)^{(p-1)p^j}G$. Consequently,

$(\text{ad } v)^{(p-1)p^j}G + G_0$ is an irreducible, transitive, depth- q graded Lie alge-

bra which is annihilated by $(\text{ad } v)^{p^j}$.

**Proof.** For $j \geq 0$ (since $(\text{ad } v)^{p^j} (\text{ad } v)^{(p-1)p^j} G = (\text{ad } v)^{p^{j+1}G=0}$, by the definition of j), we have

$$(\text{ad } v)^{(p-1)p^j} G \supseteq (\text{ad} v)^{(p-1)p^j} \left[(\text{ad} v)^{(p-1)p^j} G, G\right] = \left((\text{ad} v)^{p^j}\right)^{p-1} \left[(\text{ad} v)^{(p-1)p^j} G, G\right]$$

$$= \left[(\text{ad} v)^{(p-1)p^j} G, (\text{ad} v)^{(p-1)p^j} G\right].$$

So $\left[(\text{ad } v)^{(p-1)p^j} G, (\text{ad } v)^{(p-1)p^j} G\right] \subseteq (\text{ad } v)^{(p-1)p^j} G$; i.e. $(\text{ad } v)^{(p-1)p^j} G$ is a Lie algebra whenever $j \geq 0$, its closure under addition being obvious. Note that we must have $r \geq q(p-1)p^j - q$, since otherwise we would have

$$(\text{ad } v)^{(p-1)p^j} G \subseteq \sum_{m < -q} G_n = 0, \text{ to contradict Lemma 7.}$$

By Lemma 6 $(\text{ad } v)^{p^j} G_{p^j q} \neq 0$. By Lemma 1 (or Lemma 2), $(\text{ad } v)^{p^j} G_{p^j q}$ is a non-zero ideal of $G_0$.

Thus, by transitivity and irreducibility $\left[G_{-1}, (\text{ad } v)^{p^j} G_{p^j q}\right] = G_{-1}$. Thus, we have

$$G_{-1} = \left[G_{-1}, (\text{ad} v)^{p^j} G_{p^j q}\right] = (\text{ad} v)^{p^j} \left[G_{-1}, G_{p^j q}\right] \subseteq (\text{ad} v)^{p^j} G_{p^j q-1} \subseteq G_{-1}$$

Consequently, we conclude that $G_{-1} = (\text{ad } v)^{p^j} G_{p^j q-1}$, so $G_{-1} \subseteq (\text{ad } v)^{p^j} G$. By Lemma 6, $G_{-q} \subseteq (\text{ad } v)^{p^j} G$, also. Thus, $(\text{ad } v)^{p^j} G + G_0$ is an irreducible, transitive depth-q graded Lie algebra? Since by Lemma 7, $(\text{ad } v)^{p^j (p-1)} G \neq 0$, it follows that $(\text{ad } v)^{p^j} \left((\text{ad } v)^{p^j} G\right) \neq 0$, so we may repeat the argument just made to conclude that $\left((\text{ad } v)^{2p^j} G\right)$ is an irreducible, transitive depth-q graded Lie algebra. Repeating the argument P-3 more times, we conclude the proof of Lemma 8.

## PROOF OF MAIN THEOREM

Let $J_1$ be the maximum whole number such that $(\text{ad } v)^{p^h} G \neq 0$ for some $v \in G_{-q}$. Such a maximal $J_1$ must exist, since the height r of the finite-dimensional graded Lie algebra G is finite. If $j \geq 0$, then we are done. Suppose then that $j \geq 1$. Let $V_1$ be an element of $G_{-q}$ such that $(\text{ad } v_1)^{p^j} G \neq 0$. Then by Lemma 8, $G^{\{1\}} \overset{\text{def}}{=} (\text{ad } v_i)^{(p-1)p^{ji}} G + G_0$ is an irreducible, transitive, finite-dimensional depth-q graded Lie algebra to which we may apply Lemma 8 to obtain a $j_2$ and $v_2$ such that

$G^{\{1\}} \overset{\text{def}}{=} (\text{ad } v_i)^{(p-1)p^{j2}} (\text{ad } v_i)^{(p-1)p^{ji}} G + G_0$ is an irreducible, transitive, finite-dimensional depth-q graded Lie algebra to which we may apply Lemma 8 again. Since $G_{-q}$ is abelian, it follows that

$$(\text{ad } v_1 + \text{ad } v_2)^{p^{j1}} G^{\{1\}} = (\text{ad } v_i) G^{\{1\}} + (\text{ad } v_2)^{p^{j1}} G^{\{1\}} = (\text{ad } v_2)^{p^{j1}} G^{\{1\}}$$

Consequently, if $j_1 = j_2$, then $(\text{ad } v_2)^{pj^1} G^{\{1\}} \neq 0$, so $V_2$ is linearly independent of $V_1$. Since $G_{-q}$, like G is finite-dimensional, we can, by repeating this process, arrive at an integer $t_i \leq \dim G_{-q}$ such that $j_{t_1} = j_1$, but $j_{t_{1+m}} < j_{t_1}$ for any m>o for which $j_{t_{1-m}}$ is ultimately defined through the repetitive process we just described. Then

$$(\text{ad } v_k)^{p^{j1}} \prod_{i=1}^{t_1} \left( (\text{ad } v_i)^{(p-1)p^{ji}} \right) G = 0, \ 1 \leq k \leq t_1$$

Since the sequence $j_1, j_2, ....., j_{t_1}$ is non-increasing, the aforementioned

commutatively of $G_{-q}$ entails that $(\text{ad } G_{-1})^{p^{j1}} \prod_{i=1}^{t_1} \left( (\text{ad } v_i)^{(p-1)p^{ji}} \right) G = 0$.

If, in the above argument, we replace $v_1$ and $j_1$ with $v_{t_1+1}$ and $j_{t_1+1}$, we eventually, by the finite dimensionality of $G_{-q}$, obtain a $t_2$ such that $j_{t_2} = j_{t_1+1}$, but $j_{t_2+1} < j_{t_2}$. Continuing in this way, using $v_{t_1+1}i$ and $j_{t_1+1}i \geq 1$, in the above argument, we see that the series $j_{t_i}, i = 1, 2, \ldots$ must eventually decrease to zero; i.e., we obtain a Lie algebra $G^{\{n\}}$ such that $(\text{ad } v)^p G^{\{n\}} = 0$ for all $v \in G_{-q}$, as required.

Remark. Note that if we define, $t_0 \overset{def}{=} 0$ and $V_{ji} = < v_{t_{i+1}}, \ldots, v_{t_{i+1}} >$ for $i > 0$, and $t_0 \overset{def}{=} 0$, then we get, for depth q, something analogous to a flag in the sense of [1]

## REFERENCES

1. Kostrikin, A.I. and Shafarevich, I.R. (1969) Graded Lie Algebras of Finite Growth. Mathematics of the USSR-Izvestiya, 3, 237–304. (English).

2. Premet, A. and Strade, H. (2006) Classification of Finite-Dimensional Simple Lie Algebras in Prime Characteristics. arXiv: math/0601380v2 [math. RA].

3. Skryabin, S.M. (1992) New Series of Simple Lie Algebras of Characteristic 3. Russian Academy of Sciences Sbornik Mathematics, 70, 389–406. (English).

4. Benkart, G.M., Kostrikin, A.I. and Kuznetsov, M.I. (1996) The Simple Graded Lie Algebras of Characteristic Three with Classical Reductive Component L0. Communications in Algebra, 24, 223–234. http://dx.doi.org/10.1080/00927879608825563

5. Benkart, G.M., Gregory, T.B. and Kuznetsov, M.I. (1998) On Graded Lie Algebras of Characteristic Three with Classical Reductive Null Component. In: Ferrar, J.C. and Harada, K., Eds., the Monsteer and Lie Algebras, Vol. 7, Ohio State University Mathematical Research Publications, 149–164.

6. Gregory, T.B. and Kuznetsov, M.I. (2004) On Depth-Three Graded Lie Algebras of Characteristic Three with Classical Reductive Null Component. Communications in Algebra, 32, 3339–3371. http://dx.doi.org/10.1081/AGB-120039401

7. Gregory, T.B. and Kuznetsov, M.I. On Graded Lie Algebras of Characteristic Three with Classical Reductive Null Component. (In Preparation).

8.  Weisfeiler, B.J. (1978) On the Structure of the Minimal Ideal of Some Graded Lie Algebras in Characteristic p > 0. Journal of Algebra, 53, 344–361. http://dx.doi.org/10.1016/0021-8693(78)90280-6.

9.  Benkart, G.M. and Gregory, T.B. (1989) Graded Lie Algebras with Classical Reductive Null Component. Mathematische Annalen, 285, 85–98. http://wdx.doi.org/10.1007/BF01442673.

10. Jacobson, N. (1962) Lie Algebras, Tracts in Mathematics. Vol. 10, Interscience, New York.

## CITATION

Gregory, T. (2014) On the Initial Sub algebra of a Graded Lie Algebra. Advances in Pure Mathematics, 4, 513-517. doi: 10.4236/apm.2014.49058.

# The Localization of Commutative Bounded BCK-Algebras

## *Dana Piciu and Dan Dorin Tascau*

Faculty of Mathematics and Computer Science,
University of Craiova, Craiova, Romania

**9**

## ABSTRACT

In this paper we develop a theory of localization for bounded commutative BCK-algebras. We try to extend some results from the case of commutative Hilbert algebras (see [1]) to the case of commutative BCK-algebras.

## INTRODUCTION

In 1966, Y. Imai and K. Iséki introduced a new notion called a BCK-algebra (see [2]). This notion is originated from two different ways. One of the motivations is based on the set theory (where the set difference operation play the main role) and another motivation is from classical and non-classical propositional calculi (see [2]). There are some systems which contain the only implication functor among the logical functors. These examples are the systems of positive implicational calculus, weak positive implicational calculus by A. Church, and BCI, BCK-systems by C. A. Meredith (see [3]).

In this paper we develop a theory of localization for commutative (bounded) BCK-algebras, and then we deal with generalizations of results which are obtained in the paper [1] for case of Hilbert algebras. For some informal explanations of the theory of localization for others categories of algebras see [4, 5].

The paper is organized as follows: in Section 2 we re-call the basic definitions and put in evidence many rules of calculus in (commutative) BCK-algebras which we need in the rest of paper. In Section 3 we introduce the commutative BCK-algebra of fractions relative to a V-closed system. In Section 4 we develop a theory for multipliers on a commutative (bounded) BCK-algebra. In Section 5 we define the notions of BCK-algebras of fractions and maximal BCK-algebra of quotients for a com-mutative (bounded) BCK-algebra. In the last part of this section is proved the existence of the maximal BCK- algebra of quotients (Theorem 29). In Section 6 we develop a theory of localization for commutative (bounded) BCK-algebras. So, for commutative (bounded) BCK- algebra A we define the notion of localization BCK-algebra relative to a topology F on A. In Section 7 we describe the localization BCK-algebra $A_F$ in some special instances.

## PRELIMINARIES

In this paper the symbols $\Rightarrow$ and $\Leftrightarrow$ are used for logical implication, respectively logical equivalence.

**Definition 1:** ([6]) A BCK-algebra is an algebra $(A \rightarrow, 1)$ of type $(2, 0)$ such that the following axioms are fulfilled for every x, y, z $\in$ A:

$(a_1)$ $x \rightarrow x = 1$;

$(a_2)$ If $x \rightarrow y = y \rightarrow x = 1$, then x=y;

(B) $(x \to y) \to ((y \to z) \to (x \to z)) = 1$;

(C) $x \to (y \to z) = y \to (x \to z)$;

(K) $x \to (y \to x) = 1$.

In [7] it is proved that the system of axioms $\{a_1, a_2, B, C, \text{ and } K\}$ is equivalent with the system $\{a_2, a_3, a_4, B\}$, where:

$(a_3)$ $x \to 1 = 1$

$(a_4)$ $1 \to x = x$.

For examples of BCK-algebras see [6-8]. If A is a BCK-algebra, then the relation $\leq$ defined by $x \leq y$ iff $x \to y = 1$ is a partial order on A (which will be called the natural order on A; with respect to this order 1 is the largest element of A. A will be called bounded if A has a smallest element 0; in this case for $x \in y$ we denote $x^* = x \to 0$. If $(x \to y) \to y = (y \to x) \to x$ for every $x, y \in A$, then A is called commutative (see [5,9,10]).

We have the following rules of calculus in a BCK-algebra A (see [6, 7]):

$(c_1)$ $x \leq y \to x$;

$(c_2)$ $x \leq (x \to y) \to y$;

$(c_3)$ $((x \to y) \to y) \to y = x \to y$;

$(c_4)$ $x \to y \leq (z \to x) \to (z \to y) \leq z \to (x \to y)$;

$(c_5)$ If $x \leq y$, then for every $z \in A$, $z \to x \leq z \to y$, and $y \to z \leq x \to z$.

Proposition 1: ([9], p. 5) If A is a commutative BCK- algebra, then relative to the natural ordering, A is a join- semilattice, where for, x, y $\in$ A:

$$x \vee y = (x \rightarrow y) \rightarrow y = (y \rightarrow x) \rightarrow x.$$

Lemma 2: Let A be a commutative BCK-algebra. For every x, y, z$\in$A there exists $(x \rightarrow z) \wedge (y \rightarrow z)$ and

$(c_6)$  $(x \vee y) \rightarrow z = (x \rightarrow z) \wedge (y \rightarrow z).$

Proof: Since x, y $\leq$ x $\vee$ y by $(c_5)$ we deduce that

$$(x \vee y) \rightarrow z \leq x \rightarrow z, y \rightarrow z.$$

Let now t $\in$ A such that

$$t \leq x \rightarrow z, y \rightarrow z$$

Then x, y $\leq$ t $\rightarrow$ z $\Rightarrow$ x $\vee$ y $\leq$ t $\rightarrow$ z $\Rightarrow$ t $\leq$ (x $\vee$ y) $\rightarrow$ z that is,

$$(x \vee y) \rightarrow z = (x \rightarrow z) \wedge (y \rightarrow z).$$

In [9] (Theorem 8) and [8] (Remark 2.1.32) it is proved the following result:

Theorem 3: If A is a BCK-algebra, then the following assertions are equivalent:

1)  For every x, y, z$\in$ A,

$$x \to (y \to z) = (x \to y) \to (x \to z);$$

2) For every x, y $\in$ A,

$$x \to (x \to y) = x \to y;$$

3) For every x y $\in$ A,

$$(x \to y) \to ((y \to x) \to x) = (y \to x)$$
$$\to ((x \to y) \to y).$$

A BCK-algebra which verify one of the above equivalent conditions is called Hilbert algebra (or positive implicative BCK-algebra).

If A is a bounded BCK-algebra, we have the following rules of calculus in A (see [6]):

$(c_7)$ If $x \le y$, then $y^* \le x^*$;

$(c_8)$ $x^* = x^{***}$, $x \le x^{**}$;

$(c_9)$ $x \to y^* = y \to x^*$, $\left(x \to y^{**}\right)^{**} = x \to y^{**}$.

Remark 1: If A is a bounded commutative BCK-algebra, then for every x $\in$ A,

$$(x \to 0) \to 0 = (0 \to x) \to x \Leftrightarrow x^{**} = x,$$

that is, A is an involutive BCK-algebra (see [6], p. 115 and [9], p. 10).

For

$$x_1, \cdots, x_n, x \in A \quad (n \ge 1)$$

we will define

$$\left(x_1,\cdots,x_n;x\right)=x_1\rightarrow\left(x_2\rightarrow\cdots\left(x_n\rightarrow x\right)\cdots\right).$$

For two elements x, y $\in$ A and a natural number n $\geq$ 1 we denote

$$x\rightarrow_n y=\left(x,x,\cdots,x;y\right)$$

Where n indicates the number of occurrences of x. Clearly, if A is a Hilbert algebra, then $x|\rightarrow_n y=x\rightarrow y$, for every n $\geq$ 1.

Let A be a BCK-algebra. A deductive system (or i-filter) of A is a non-empty subset D of A such that 1 $\in$ D and for every x, y $\in$ A, if x, x$\rightarrow$y$\in$ D, then y $\in$ D. It is clear that if D is a deductive system, x $\leq$ y and x $\in$ D then y $\in$ D We denote by Ds (A) the set of all deductive systems of A. For a nonempty subset X $\subseteq$ D, we denote by

$$\langle X\rangle=\cap\{D\in Ds\left(A\right):X\subseteq D\}$$

($\langle X\rangle$ is called the deductive system generated by X) .It is known that

$$\langle X\rangle=\{x\in A:\left(x_1,\cdots,x_n;x\right)=1,\text{ for some }x_1,\cdots,x_n\in X\}.$$

In particular for a $\in$ A, we denote by $\langle a\rangle$ the deductive system generated by $\{a\}$ ($\langle a\rangle$ is called principal and $\langle a\rangle=\{x\in A:a\rightarrow_n x=1,\text{ for some n}\geq 1\}$).

Lemma 4: Let A be a bounded BCK-algebra and x,y $\in$ A such that there exists x $\vee$ y in A. Then there exists $x^*\wedge y^*$ $x^*\wedge y^*=(x\vee y)^*$.

Proof: Clearly, $(x \vee y)^* \leq x^*, y^*$. Let $t \in A$ such that $t \leq x^*, y^*$ Then $x, y \leq t^* \Rightarrow x \vee y \leq t^* \Rightarrow t^{**} \leq (x \vee y)^*$. From $(c_8)$ we de- duce that $t \leq t^{**} \leq (x \vee y)^* \Rightarrow t \leq (x \vee y)^*$, that is,

$$\left(x \vee y\right)^* = x^* \wedge y^*$$

**Definition 2:** ([7], p. 944) Let A be a bounded BCK- algebra. An element $x \in A$ is called boolean if

$$< x > \cap < x^* > = \{1\} \quad \text{(clearly, } \langle x \rangle \cup \langle x^* \rangle = A).$$

We denote by B (A) the set of all boolean elements of A; clearly, 0, 1 ∈ B (A).

**Lemma 5:** ([7]) Let A be a BCK-algebra. Then for every

$$x, y \in A, \quad x \vee y = 1 \Leftrightarrow < x > \cap < y > = \{1\}.$$

**Corollary 6:** For a bounded BCK-algebra $x \in B(A)$ iff $x \vee x^* = 1$.

**Remark 2:** If $x \in B$ (A) that is, $x \vee x^* = 1$, then from Lemma 4 we deduce that

$$x^* \wedge x^{**} = \left(x \vee x^*\right)^* = 1^* = 0,$$

Hence $x \wedge x^* \leq x^{**} \wedge x^* = 0 \Rightarrow x \wedge x^* = 0$, that is, x* is the complement of x in A.

Boolean elements also satisfy several interesting properties which can be proved using above corollary and some arithmetical calculus:

Proposition 7: ([7]) Let A be a bounded BCK-algebra. Then for every a $\in$ B (A) and, x, y $\in$ A we have:

$(c_{10})$ $a^* \in B(A)$;
$(c_{11})$ $a \to (a \to x) = a \to x$;
$(c_{12})$ $a \to (x \to y) = (a \to x) \to (a \to y)$;
$(c_{13})$ $a \to a^* = a^*, a^* \to a = a$;
$(c_{14})$ $a^{**} = a$;
$(c_{15})$ $(a \to x) \to a = a$;
$(c_{16})$ $(a \to x) \to x \leq (x \to a) \to a$;
$(c_{17})$ $((a \to x) \to a^*) \to a^* = a \to x^{**}$;
$(c_{18})$ If $b \in B(A)$, then $(a \to b) \to b = (b \to a) \to a$;
$(c_{19})$ $a^* \to x = a \vee x = (a \to x) \to x$;
$(c_{20})$ $(a \to x^*)^* = a \wedge x^{**}$.

Corollary 8: ([7]) Let A be a bounded BCK-algebra. Then

1) If a $\in$ B(A) then

$$\langle a \rangle = [a] = \{x \in A : a \leq x\};$$

2) For a, b $\in$ B(A)

$$a \to b \in B(A);$$

3) $(B(A), \to, 0, 1)$ is a Boolean algebra (where for

$a, b \in B(A), a \vee b = a^* \to b$ and $a \wedge b = (a \to b^*)^*$

Corollary 9: Let A be a commutative BCK-algebra. For every $a \in B(A)$ and $y, z \in A$ we have:

$(c_{21})$ $a \vee (y \rightarrow z) = (a \vee y) \rightarrow (a \vee z)$.

Proof: By $(c_6)$ we have

$$(a \vee y) \rightarrow (a \vee z) = (a \rightarrow (a \vee z)) \wedge (y \rightarrow (a \vee z))$$
$$= 1 \wedge (y \rightarrow (a \vee z)) = y \rightarrow (a \vee z) = y \rightarrow ((a \rightarrow z) \rightarrow z)$$

So $(c_{21})$ is equivalent with $(*) a \vee (y \rightarrow z) = y \rightarrow (a \vee z)$. Clearly,

$$a \leq a \vee z \leq y \rightarrow (a \vee z)$$

and from

$$z \leq a \vee z \Rightarrow y \rightarrow z \leq y \rightarrow (a \vee z).$$

So to prove (*) let $t \in A$ such that $a \leq t$ and $y \rightarrow z \leq t$. we have the intention to prove that

$$y \rightarrow (a \vee z) \leq t \Leftrightarrow y \rightarrow ((a \rightarrow z) \rightarrow z) \leq t$$
$$\Leftrightarrow (**)(a \rightarrow z) \rightarrow (y \rightarrow z) \leq t.$$

Indeed, from

$$y \to z \leq t \Rightarrow (a \to z) \to (y \to z)$$

$$\overset{(B)}{\leq} (a \to z) \to t \leq (t \to a) \to \left[(a \to z) \to a\right]$$

$$\overset{(c_{15})}{=} (t \to a) \to a = (a \to t) \to t = 1 \to t = t.$$

Proposition 10: Let A be a commutative BCK-algebra. Then for every $a,b \in B(A)$ and $x \in A$ we have:

$(c_{22})$ $(a \vee x) \to (b \vee x) = (a \to b) \vee x.$

Proof: By $(c_6)$ we have

$$(a \vee x) \to (b \vee x)$$

$$= \left[a \to (b \vee x)\right] \wedge \left[x \to (b \vee x)\right]$$

$$= \left[a \to (b \vee x)\right] \wedge 1 = a \to ((x \to b) \to b).$$

Also

$$(a \to b) \vee x = (x \to (a \to b)) \to (a \to b)$$

$$\overset{(C)}{=} (a \to (x \to b)) \to (a \to b)$$

$$\overset{(c_{12})}{=} a \to ((x \to b) \to b).$$

**Definition 3:** If $A_1$ $A_2$ are BCK-algebras, then $f : A_1 \to A_2$ is called morphism of BCK-algebras if $f(x \to y) = f(x) \to f(y)$, for every $x,y \in A_1$ (if $A_1$ $A_2$ are bounded BCK-algebras, then we add the condition $f(0) = 0$).

## COMMUTATIVE BCK-ALGEBRA OF FRACTIONS RELATIVE TO A V-CLOSED SYSTEM

In this section by A we denote a commutative bounded BCK-algebra.

**Definition 4:** A nonempty subset S of A will be called V-closed system of A if $0 \in S$ and $x \vee y \in S$ for every $x, y \in S$.

For a V-closed system $S \subseteq A$ we define the binary relation $\theta_s$ on A by $(x, y) \in \theta_s$ if there is $s \in S \cap B(A)$ such that $s \vee x = s \vee y$.

Proposition 11: The relation $\theta_s$ is a congruence on A.

Proof: Clearly $\theta_s$ is an equivalence relation on A. To prove the compatibility of $\theta_s$ with the operation $\to$, let $x, y, z \in A$ such that $(x, y) \in \theta_s$ (hence there is $s \in S \cap B(A)$ such that $s \vee x = s \vee y$ By $(c_{21})$ we deduce

$$s \vee (z \to x) = (s \vee z) \to (s \vee x)$$
$$= (s \vee z) \to (s \vee y) = s \vee (z \to y),$$

and similarly,

$$s \vee (x \to z) = s \vee (y \to z),$$

that is,

$$(z \to x, z \to y) \in \theta_S$$

and

$(x \rightarrow z, y \rightarrow z) \in \theta_S$.

We denote $A[S] = A/\theta_s$; the commutative BCK- algebra $A[S]$ will be called BCK-algebra of fractions of A relative to S. For $x \in A$ we denote by $[x]_{\theta_s}$ the equivalence class of x relative to $\theta_s$. Clearly, in

$A[s], 1 = [1]_{\theta s} = \{x \in A : (x,1) \in \theta s = \{x \in A : \text{there is } s \in S \cap B(A) \text{ such that } s \vee x = 1\}$,

$0 = [0]\theta_s = \{x \in A : (x,0) \in \theta_s\} = \{x \in A : \text{there is } s \in S \cap B(A) \text{ such that } s \vee x = s\}$

$= \{x \in A : \text{there is } s \in S \cap B(A) \text{ such that } x \leq s\}$ and for every $x, y \in A, [x]_{\theta s} \rightarrow [y]_{\theta s} = [x \rightarrow y]_{\theta s}$.

**Proposition 12:** A[s] is a bounded commutative BCK- algebra, when $0 = [S]_{\theta s}$ with $s \in S \cap B(A)$.

Proof:    Clearly,    if $s, t \in S \cap B(A)$ since    $r = s \vee t \in S \cap B(A)$ and $r \vee s = r \vee t \Rightarrow [s]_{\theta s} = [t]_{\theta s}$. To prove that, for $s \in S \cap B(A), [s]_{\theta s} = 0$ let $x \in A$. We have

$$[s]_{\theta_S} \leq [x]_{\theta_S} \Leftrightarrow [s]_{\theta_S} \vee [x]_{\theta_S} = [x]_{\theta_S} \Leftrightarrow [s \vee x]_{\theta_S} = [x]_{\theta_S}$$

Which is true since $s \vee (s \vee x) = s \vee x$.

We denote by $P_s : A \rightarrow A[S]$ the canonical subjective morphism of BCK-algebras (defined by $P_s(x) = [x]_{\theta s}$ for every $x \in A$ ).

Remark 3: since for every $s \in S \cap B(A)$,$\}$ $s \vee s = s \vee 0$ we deduce that $P_s(S \cap B(A)) = \{0\}$.

Proposition 13: If $x \in A$, then $[x]_{\theta_S} \in B(A[S])$ iff there exists $s \in S \cap B(A)$ such that $x \vee x^* \vee s = 1$ so, if $x \in B(A)$, than $[x]_{\theta_S} \in B(A[S])$.

Proof: For $x \in A$, we have

$$[x]_{\theta_S} \in B(A[S]) \Leftrightarrow [x]_{\theta_S} \vee \left([x]_{\theta_S}\right)^*$$
$$= 1 \Leftrightarrow \left[x \vee x^*\right]_{\theta_S} = 1 \Leftrightarrow$$

there exists $s \in S \cap B(A)$ such that $x \vee x^* \vee s = 1 \vee s = 1$ If $x \in B(A)$, since $x \vee x^* \vee 0 = 1$ and $0 \in S \cap B(A)$, we deduce that $[x]_{\theta_S} \in B(A[S])$.

A[S] Verify the following property of universality:

Theorem 14: For every bounded commutative BCK- algebra B and every morphism of bounded BCK-algebras $f : A \to B$ such that $f(S \cap B(A)) = \{0\}$, there exists a unique morphism of bounded BCK-algebras $f' : A[S] \to B$ such that $f' \circ p_s = f$.

Proof: Let $x, y \in A$ such that $[x]_{\theta_S} = [y]_{\theta_S}$. Then there is $s \in S \cap B(A)$ such that

$$s \vee x = s \vee y \Rightarrow f(s \vee x) = f(s \vee y)$$
$$\Rightarrow f(s) \vee f(x) = f(s) \vee f(y)$$
$$\Rightarrow 0 \vee f(x) = 0 \vee f(y) \Rightarrow f(x) = f(y).$$

So, $f': A[S] \rightarrow B$ defined for $x \in A$ by $f'([x]_{\theta s}) = f(x)$ is correct defined. Clearly, $f'$ is morphism of bounded BCK-algebras and $f' \circ p_s = f$. The unity of $f'$ follows from the fact that $P_s$ is onto.

Example 1: If A is a bounded commutative BCK-algebra and S{0} or S such that $0 \in S$ and $S \cap B(A) \setminus \{0\}) = \varnothing$ then for $x, y \in A, (X, y) \in \theta s \Leftrightarrow x \vee 0 = y \vee 0 \Leftrightarrow x = y$, hence A[S] = A.

Example 2: If A is a bounded commutative BCK-algebra and S is an V-closed system system such that $1 \in S$ (for example S=A or S=B (A)) then for every $x, y \in A, (X, y) \in \theta s$ (Since $x \vee 1 = y \vee 1$ and $1 \in S \cap B(A)$ hence in this case A[S] = {1}.

**Definition 5:** A[S] is called the BCK-algebra of fractions of A relative to S.

## MULTIPLIERS ON A COMMUTATIVE BOUNDED BCK-ALGEBRA

The concept of maximal lattice of quotients for a distributive lattice was defined by J. Schmid in [11,12] (taking as a guide-line the construction of complete ring of quotients by partial morphisms introduced by G. Findlay and J. Lambek (see [13], p. 36). The central role in the construction of the maximal lattice of quotients for a distributive lattice due to J. Schmidt in [11] and [12] is played by the concept of multiplier for a distributive lat-tice defined by W. H. Cornish in [14].

In this section we develop a theory for multipliers on a commutative bounded BCK-algebra A.

**Definition 6:** A subset $T \subseteq A$ is called V-subset A if for every $a \in A$ and $x \in T$ we have $a \vee x \in T$.

We denote by T (A) the set of all V-subsets of A. Clearly $Ds(A) \subseteq T(A)$ (and more generally, if denote by I (A) the set of all increasing subsets of A, then $I(A) \subseteq T(A)$).

Remark 4: Clearly, if $D_1, D_2 \in T(A)$ then $D_1 \cap D_2 \in T(A)$.

Lemma 15: If $D \in T(A)$, then

1.  $1 \in D$

2.  If $X \leq y$ and $x \in D$ then $y \in D$

Proof: (i). If $x \in D$ since $1 \in A$ then $1 = 1 \vee x \in D$ 3) we have $y = y \vee x$.

**Definition 7:** By partial strong multiplier on A we mean a map $f : D \rightarrow A$ where $D \in T(A)$, such that:

$(sm_1)$ For every $x \in D$ and $e \in B(A)$.

$$f(e \vee x) = e \vee f(x);$$

$(sm_2)$ For every $x \in D$ $x \leq f(x)$;

$(sm_3)$ If $e \in D \cap B(A)$, then $f(e) \in B(A)$;

(sm$_4$) For every $x \in D$ and $e \in D \cap B(A)$,

$$f(e) \vee x = e \vee f(x).$$

By $\mathrm{dom}(f) \in T(A)$ we denote the domain of f; if dom (f) =A, we called f total.

To simplify the language, we will use strong multiplier instead partial strong multiplier using total to indicate that the domain of a certain multiplier is A.

Examples

1. The map $0,1: A \to A$ defined by $0(x) = x$ and respectively $1(x) = 1$, for every $x \in A$ are total strong multiplier on A.

2. For $a \in B$ and $D \in T(A)$, the map $f_a : D \to A$ defined by $f_a(x) = a \vee x$, $X \in D$ is a strong multiplier on A (called principal).

If dom $(f_a) = A$, we denote $f_a$ by $\overline{f_a}$.

Remark 5: If $f : D \to A$ is a strong multiplier on A (with $D \in T(A)$), then $f(1) = 1$ Indeed, if in (sm$_1$) we put e=1, we obtain that for every $x \in D$,

$$f(1 \vee x) = 1 \vee f(x) \Leftrightarrow f(1) = 1.$$

For $D \in T(A)$, we denote $M(D,A) = \{f : D \to A : f \text{ is a strong multipier on } A\}$

and $M(A) = \bigcup_{D \in T(A)} M(D,A)$.

For $D_1, D_2 \in T(A)$ and $f_i \in M(D_i, A), i = 1, 2$, we define $f_1 \rightarrow f_2 : D_1 \cap D_2$ by $(f_1 \rightarrow f_2)(x) = f_1(x) \rightarrow f_2(x)$, for every $x \in D_1 \cap D_2$.

Lemma 16: $f_1 \rightarrow f_2 \in M(D_1 \cap D_2, A)$.

Proof: If $x \in D_1 \cap D_2$ and $e \in B(A)$ then

$$(f_1 \rightarrow f_2)(e \vee x) = f_1(e \vee x) \rightarrow f_2(e \vee x)$$
$$= (e \vee f_1(x)) \rightarrow (e \vee f_2(x)) \overset{(c_{21})}{=} e \vee (f_1(x) \rightarrow f_2(x))$$
$$= e \vee (f_1 \rightarrow f_2)(x),$$

$$(f_1 \rightarrow f_2)(x) = f_1(x) \rightarrow f_2(x) \geq f_2(x) \overset{(sm_2)}{\geq} x,$$
$$(f_1 \rightarrow f_2)(e) = f_1(e) \rightarrow f_2(e) \in B(A)$$

By Corollary 8 (since $f_1(e), f_2(e) \in B(A)$ and if

$$e \in D_1 \cap D_2 \cap B(A),$$
$$e \vee (f_1 \rightarrow f_2)(x) = e \vee (f_1(x) \rightarrow f_2(x))$$
$$\overset{(c_{21})}{=} (e \vee f_1(x)) \rightarrow (e \vee f_2(x))$$
$$\overset{(sm_4)}{=} (x \vee f_1(e)) \rightarrow (x \vee f_2(e))$$
$$\overset{(c_{22})}{=} x \vee (f_1(e) \rightarrow f_2(e)) = x \vee (f_1 \rightarrow f_2)(e),$$

That is $f_1 \rightarrow f_2(e) \in M(D_1 \cap D_1, A)$.

Corollary 17: $(M(A), \rightarrow 0, 1)$ is a bounded commutative Back-algebra.

Proof: The fact that M (A) is a commutative BCK-algebra follows from Lemma 16. If $f \in M(D, A)$ and $x \in D$, then $0(x) \le x \le f(x) \le 1 \le 1(x)$ and since the relation of order on M (A) is given by $f_1 \le f_2$ iff $f_1(x) \le f_2(x)$ for every $x \in dom(f_1) \cap dom(f_2)$ we deduce that $0 \le f \le 1$ that is, M (A) is bounded.

Lemma 18: The map $v_A(a) : B(A) \rightarrow M(A)$ is defined by $v_A(a) = \overline{f_a}$ for every $a \in B(A)$ is a morphism of bounded BAK-algebras.

Proof: If $a, b \in B(A)$ and $x \in A$, then

$$\left( \overline{f_a} \rightarrow \overline{f_b} \right)(x) = \overline{f_a}(x) \rightarrow \overline{f_b}(x)$$

$$= (a \vee x) \rightarrow (b \vee x) \overset{(c_{22})}{=} (a \vee b) \rightarrow x = \overline{f_{a \rightarrow b}}(x),$$

So, $v_A(a) \rightarrow v_A(b) = v_A(a \rightarrow b)$ and

$$v_A(0) = \overline{f_0} = 0.$$

**Definition 8:** $D \subseteq A$ is called regular if for every $x, y \in A$ such that $e \vee x = e \vee y$ for every $e \in D \cap B(A)$, then x=y.

**For example a bounded for exam BCK-algebra is regular since** $D \in R(A)$ **we deduce that** $(f \vee f^*)(x) = 1(x)$, **hence** $f \vee f^* = 1$, **that is, Mr (A) is a Boolean algebra (by Corollary 6).**

Remark 6: The axioms $sm_3$, $sm_4$ were necessary in the proof of proposition 21.

**Definition 9:** Given two strong multipliers $f_1$, $f_2$ on A, since if $x, y \in A$ such that $e \vee x = e \vee y$ for every $e \in A \cap B(A) = B(A)$, then in particular, for e=0 we obtain $x \vee 0 = y \vee 0 \Rightarrow x = y$.

If A is, $D \in T(A)$ and $0 \in D$, then D is regular. We denote

$$R(A) = \{D \subseteq A : D \text{ is a regular subset of } A\}.$$

Lemma 19: If $D_1, D_2 \in T(A) \cap R(A)$ then $D_1 \cap D_2 \in T(A) \cap R(A)$.

Proof: By Remark 4, $D_1 \cap D_2 \in T(A)$. Let $x, y \in A$ such that $e \vee x = e \vee y$ for every.

For every $e \in D_1 \cap D_2 \cap B(A), i = 1, 2$, since $e_1 \vee e_2 \in D_1 \cap D_2 \cap B(A)$ we have

$$(e_1 \vee e_2) \vee x = (e_1 \vee e_2) \vee y \Rightarrow e_1 \vee (e_2 \vee x)$$
$$= e_1 \vee (e_2 \vee y) \Rightarrow e_2 \vee x = e_2 \vee y \Rightarrow x = y,$$

So,

$$D_1 \cap D_2 \in R(A).$$

We denote

$$M_r(A) = \{f \in M(A) : dom(f) \in T(A) \cap R(A)\}.$$

Corollary 20: $M_r(A)$ is a BCK-subalgebra of $M(A)$.

Proposition 21: $M_r(A)$ is a Boolean subalgebra of $M(A)$.

Proof: Let $f : D \to A$ be a strong multiplier on A with $D \in T(A) \cap R(A)$. Then $f^* : D \to A$, $f^*(x) = (f \to 0)(x) \to x$ for $x \in D$. We have

$$\left(f \vee f^*\right)(x) = f(x) \vee \left(f(x) \to x\right)$$
$$= \left[\left(f(x) \to x\right) \to f(x)\right] \to f(x)$$

Then for $e \in D \cap B(A)$ and $x \in D$ we have

$$e \vee \left[f \vee f^*\right](x) = e \vee \left[\left[\left(f(x) \to x\right) \to f(x)\right] \to f(x)\right]$$
$$\overset{(c_{21})}{=} \left[\left[\left(\left(e \vee f(x)\right) \to (e \vee x)\right) \to \left(e \vee f(x)\right)\right] \to \left(e \vee f(x)\right)\right]$$
$$\overset{(sm_4)}{=} \left[\left[\left(\left(x \vee f(e)\right) \to (e \vee x)\right) \to \left(x \vee f(e)\right)\right] \to \left(x \vee f(e)\right)\right]$$
$$\overset{(c_{22})}{=} x \vee \left[\left[\left(f(e) \to e\right) \to f(e)\right] \to f(e)\right] = x \vee \left[\left(f(e) \to e\right) \vee f(e)\right]$$
$$= x \vee \left[\left(f(e)\right)^* \vee e \vee f(e)\right] = x \vee 1 = 1 = e \vee 1 = e \vee 1(x).$$

We say that $f_1$ extends $f_2$ if $\text{dom}(f_2) \subseteq \text{dom}(f_1)$ and $f_1(x) = f_2(x)$, for all $x \in \text{dom}(f_2)$; we write $f_2 \leq f_1$ if $f_1$ extends $f_2$. A strong mtultiplier $f$ is called maximal if $f$ cannot be extend to a strictly larger domain.

**Lemma 22:**

1. If $f_1, f_2 \in M(A), f \in M_r(A)$ and $f < f_1, f < f_2$, then $f_1$ and $f_2$ coincide on the $\text{dom}(f_1) \cap \text{dom}(f_2)$;

2. Every strong multiplier $f \in M_r(A)$ can be extended to a maximal strong multiplier $f_a$. More precisely, each principal strong multiplier $f_a$ with $a \in B(A)$ and $\text{dom}(f_a) \in T(A) \cap R(A)$ can be uniquely extended to the total strong multiplier $\overline{f}_a$ and each non-principal strong multiplier can be extended to a maximal non-principal one.

**Proof:**

1. If by contrary, there exists $t \in \text{dom}(f_1) \cap \text{dom}(f_2)$ such that $f_1(t) \neq f_2(t)$, since $\text{dom}(f) \in R(A)$, then there exist $t' \in \text{dom}(f) \cap B(A)$ such that $t' \vee f_1(t) \neq t' \vee f_2(t) \Leftrightarrow f_1(t' \vee t) \neq f_2(t' \vee t)$ which is contradictory, since $t' \vee t \in \text{dom}(f)$.

2. We first prove that $f_a$ with $a \in B(A)$ cannot be extended to a non-principal strong multiplier. Let $D = \text{dom}(f_a) \in T(A) \cap R(A), f_a : D \to A$ and suppose by contrary that there exist $D' \in T(A), D \subseteq D'$, (hence $D' \in T(A) \cap R(A)$ ) and a non-principal strong multiplier $f \in M(D', A)$ which extends $f_a$. Since $f$ is non-principal, there exists $x_0 \in D' \in x_0 \notin D$ such that $f(x_0) \neq a \vee x_0$. Since $D \in R(A)$, then there exists $t \in D \cap B(A)$ such that

$$t \vee f(x_0) \neq t \vee (a \vee x_0) \Leftrightarrow f(t \vee x_0) \neq a \vee (t \vee x_0),$$

Which is contradictory since $f_a \leq f$. Hence $f_a$ is uniquely extended by $\overline{f_a}$.

Now, let $f \in M_r(A)$ be non-principal and $M_f = \{(D,g) : D \in T(A), g \in M(D,A), \mathrm{dom}(f) \subseteq D \text{ and } g_{|\mathrm{dom}(f)} = f\}$ (clearly if $(D,g) \in M_f$, then $(D,g) \in M_f D \in T(A) \cap R(A)$).

The set $M_f$ is ordered by $(D_1, g_1) \leq (D_2, g_2)$ iff $D_1 \subseteq D_2$ and $g_{2|D_1} = g_1$. Let $(D_k, g_k) : k \in K$ be a chain in $M_f$. Then $D' = \bigcup_{k \in K} D_k \in T(A)$ and $\mathrm{dom}(f) \subseteq D'$. So, $g' : D \to A$ defined by $g'(x) = g_k{}'(x)$ if $x \in D_k$ is correctly defined (since if $x \in D_k \cap D_t$ with $k, t \in K$, then by 1), $g'(x) = g_k{}'(x)$).

Clearly $g' \in M(D', A)$, and $g'_{|\mathrm{dom}(f)} = f$ (since if $x \in \mathrm{dom}(f) \subseteq D'$ then $x \in D'$ and so there exist $k \in K$, such that $x \in D_k$, hence $g'(x) = g_k(x) = f(x)$).

So, $(D', g')$ is an upper bound for the family $\{(D_k, g_k : k \in K)\}$, hence by Zorn's lemma, $M_f$ contains at least one maximal strong multiplier $h$ which extends f. Since f is non-principal and h extends f, h is also non-principal.

On the Boolean algebra $M_r(A)$ we consider the relation $\rho_A$ defied by $(f_1, f_1) \in \rho_A$ iff $f_1$ and $f_2$ conside on the intersection of their domains.

Lemma 23: $\rho_A$ is a congruence on $M_r(A)$.

Proof: The reflexivity and the symmetry of $\rho_A$ are immediately; to prove the transitivity of $\rho_A$ let $(f_1, f_2), (f_2, f_3) \in \rho_A$. Therefore $f_1, f_2$, and respectively $f_2, f_3$ coincide on the intersection of their domain. If by contrary, there exists $x_0 \in \text{dom}(f_1) \cap \text{dom}(f_3)$ such that $f_1(x_0) \neq f_3(x_0)$ since $\text{dom}(f_2) \in \text{dom}(f_3)$, there exists $t \in \text{dom}(f_2) \cap B(A)$ such that

$$t \vee f_1(x_0) \neq t \vee f_3(x_0) \Leftrightarrow f_1(t \vee x_0) \neq f_3(t \vee x_0)$$

Which is contradictory, since $t \vee x_0 \in \text{dom}(f_1) \cap \text{dom}(f_2) t \cap \text{dom}(f_3)$. The compatibility of $\rho_A$ with $\rightarrow$ on $M_r(A)$ is immediately.

For $f \in M_r(A)$ we denote by [f] the congruence class of f modulo $\rho_A$ and $A'' = M_r(A) / \rho_A$.

Remark 7: From proposition 21 wen deduce that A'' is a Boolean algebra.

Lemma 24: The map $\overline{v_A} : B(A) \rightarrow A''$ defined by $\overline{v_A}(a) = [\overline{f_a}]$ is an injective morphism of Boolean algebras and $\overline{v_A}(B(A)) \in R(A'')$.

Proof: The fact that $\overline{v_A}$ is a morphism of Boolean algebras follows from Lemma 18. To prove the injectivity of $\overline{v_A}$ let $a, b \in B(A)$ such that $\overline{v_A}(a) = \overline{v_A}(b)$. Then

$$\left[\overline{f_a}\right] = \left[\overline{f_b}\right] \Leftrightarrow \left(\overline{f_a}, \overline{f_b}\right) \in \rho_A \Leftrightarrow \overline{f_a}(x) = \overline{f_b}(x),$$

or every $x \in A \Leftrightarrow x \vee a = x \vee b$, for every, hence for x=0 we obtain that $0 \vee a = 0 \vee b \Rightarrow a = b$. To prove $\overline{v_A}(B(A)) \in R(A)$, if by contrary there exist such that $[f_1] \neq [f_2]$ (that is there exist $x_0 \in$ dom $(f_1) \cap$ dom$(f_2)$ such that $f_1(x_0) = f_2(x_0)$ and

$$\left[ f_1 \right] \vee \left[ \overline{f_a} \right] = \left[ f_2 \right] \vee \left[ \overline{f_a} \right] \Leftrightarrow \left[ f_1 \vee \overline{f_a} \right] = \left[ f_2 \vee \overline{f_a} \right]$$

for every $f_1(x) \vee a \vee x = f_2(x) \vee a \vee x$, for every $a \in B(A)$ and every $x \in$ dom $(f_1) \cap$ dom$(f_2)$. For $a = 0$ and $x = x_0$ we obtain that

$$f_1(x_0) \vee x_0 = f_2(x_0) \vee x_0 \Leftrightarrow f_1(x_0) = f_2(x_0)$$

Which is contradictory.

Remark 8: Since for every $a \in B$ (A), $\overline{f_a}$ is the unique maximal strong multiplier on $[\overline{f_a}]$ (by Lemma 22) we can identify $[\overline{f_a}]$ with $\overline{f_a}$ So, since $\vee_A$ is injective moiphism of Boolean algebras, the elements of B (A) can be identified with the elements of the set $\{\overline{f_a} : a \in B(A)\}$.

Lemma 25: In view of the identifications made above, if $[f] \in A''$ de (with $f \in M_r(A)$ and $D =$ dom$(f) \in T$ (A) $\cap R(A)$), then

$$D \cap B(A) \subseteq \{a \in B(A) : \overline{f_a} \vee [f] \in B(A)\}.$$

Proof: Let $a \in D \cap B(A)$ If by contrary, $\overline{f_a} \vee [f] \in B$ then $\overline{f_a} \vee f$ is a non-principal strong multiplier. Then by Lemma 22, (2), $\overline{f_a} \vee f$ can be extended to a non-principal maximal strong multiplier $\overline{f} : \overline{D} \to A$ with $\overline{D} \to T(A)$. Thus, $D \subseteq \overline{D}$ and for every $x \in D$,

$$\overline{f}(x) = \left(\overline{f_a} \vee f\right)(x) = a \vee x \vee f(x) = a \vee f(x).$$

Since $a \in D \cap B(A)$, then $\overline{f}(x) = f(a \vee x) \overset{(sm_4)}{=} x \vee f(a)$ that is $\overline{f}_{|D}$, is principal which is contradictory with the assumption that $\overline{f}$ is non-principal.

## MAXIMAL COMMUTATIVE BCK-ALGEBRA OF QUOTIENTS

The goal of this section is to define (taking as a guide-line the case of distributive lattices) the notions of BCK-algebra of fractions and maximal BCK-algebra of quotients for a commutative bounded BCK-algebra. For some informal explanations of notions of fraction see [13] and [5].

**Definition 10:** A bounded commutative BCK-algebra A′ is called BCK-algebra of fractions of A if:

$(f_1)$ B (A) is a BCK-subalgebra of A′;

$(f_2)$ For every $a', b', c' \in A', a' \neq b'$, there exists $a \in B$ (A) such that $a \vee a' = a \vee b'$ and $a \vee c' \in B(A)$.

As a notational convenience, we write $A \succ A'$ to indicate that A′ is a BCK-algebra of fractions of A. So, B (A) $\succ$ B (A) (since for $a', b', c' \in B(A)$ with $a' \neq b'$, if consider $0 \in B$ (A), then $a' = a' \vee 0 \neq b' \vee 0 = b'$ and $c' = c' \vee 0 \in B(A)$).

**Definition 11:** Q (A) is the maximal (commutative) BCK-algebra of quotients of A if $A \succ Q$ (A) and for every commutative and bounded BCK-algebra A′ with $A \succ A'$, there exists a monomorphism of BACK-algebras i: $A' \rightarrow Q$ (A).

Proposition 26: Let A be a commutative and bounded BCK-algebra such that $A \succ A'$. Then A' is a Boolean algebra.

Proof: If by contrary, A' is not a Boolean algebra, then by Corollary 6, there exists $x \in A'$ such that $x \vee x^* \neq 1$. Since $A \succ A'$, then there exists e $\in B(A)$, such that $e \vee x \in B(A)$ and $e \vee (x \vee x^*) \neq e \vee 1 = 1$.

Then, by Lemma 4,

$$\left(e \vee x\right) \vee \left(e \vee x\right)^* = 1 \Rightarrow \left(e \vee x\right) \vee \left(e^* \wedge x^*\right) = 1$$
$$\Rightarrow 1 \leq \left(e \vee x \vee e^*\right) \wedge \left(e \vee x \vee x^*\right)$$
$$\Rightarrow 1 \leq 1 \wedge \left(e \vee x \vee x^*\right) \Rightarrow e \vee x \vee x^* = 1,$$

a contradiction!.

Remark 9: If A is a. Boolean algebra, then $B(A) = A$. By Proposition 26, $Q(A)$ is a Boolean algebra and the axioms $sm_1$-$sm_4$ are equivalent with $sm_1$, hence $Q(A)$ is in this case just the classical Dedekind-MacNeille completion of A (see [12], p. 687). In contrast to the general situation, the Dedekind-MacNeille completion of a Boolean algebra is again dis-tributive and, in fact, is a Boolean algebra (see [15], p. 239).

Lemma 27: Let $A \succ A'$; then for every $a', b' \in A', a' \neq b'$ and any finite sequence $c'_1, \ldots, c'_n \in A'$, there exists $a \in B(A)$ such that $a \vee a' \neq a \vee b'$ and $a \vee c'_i \in B(A)$ for $i = 1, 2, ---, n$ $(n \geq 2)$.

Proof: Assume lemma holds true for n-1. So we may find $b \in B(A)$ such that $b \vee a' \neq b \vee b'$ and $b \vee c'_i \in B(A)$ for $i = 1, 2, ---, n-1$. Since $A \succ A'$, we find $c \in B(A)$ such that $c \vee (b \vee a') \neq c \vee (b \vee b')$ and. $c \vee c'_n \in B(A)$ The element $a = b \vee c \in B(A)$ has the required properties.

Lemma 28: Let $A \succ A'$ and $a' \in A'$. Then

$$D_{a'} = \{a \in B(A) : a \vee a' \in B(A)\} \in T(B(A)) \cap R(A).$$

Proof: If $a \in B$ $(A)$ and $x \in D_{a'}$, then $x \vee a' \in B$ $(A)$ and since $(a \vee x) \vee a' = a \vee (x \vee a) \in B(A)$ it follows $a \vee x \in D_{a'}$, hence $D_{a'} \in T(B(A))$. To prove $D_{a'} \in R$ $(A)$ consider $x, y \in A$ such that $e \vee x = e \vee y$, for every $e \in D_{a'} \cap B(A)$. If by contrary, $x \neq y$, since $A \succ A'$, there exists $a_0 \in B(A)$ such that $a_0 \vee a' \in B(A)$ (that is; $a_0 \in D_{a'}$) and $a_0 \vee x \neq a_0 \vee y$, which is contradictory.

Theorem 29: $A''$ (defined in section 4) is the maximal (commutative) BACK-algebra of quotients Q (A) of A.

Proof: The fact that B(A) is BACK-sub algebra (Boolean subalgebra) of Q(A) follows from Lemma 24 and remark 8. To prove $A \succ Q(A)$, $[f], [g], [h] \in Q(A)$ with $f, g, h \in M_r(A)$ such that $[g] \neq [h]$ (that is, there exists $x_0 \in \text{dom}(g) \cap \text{dom}(h)$ such that $g(x_0) \neq h(x_0)$).

Put D $=\text{dom}$ (f) $\in T(A) \cap R(A)$ and

$$D_{[f]} = \{a \in B(A) : \overline{f_a} \vee [f] \in B(A)\}.$$

Then by Lemma 25, $D \cap B(A) \subseteq D_{[f]}$ If suppose that for every a $a \in D \cap B(A), \overline{f_a} \vee [g] = \overline{f_a} \vee [h]$ then $\overline{f_a} \vee [g] = \overline{f_a} \vee [h]$, hence for every $x \in$

dom (g) $\cap$ dom(h) we have $(\bar{f_a} \vee g)(x) = (\bar{f_a} \vee h)(x) \Leftrightarrow$ (analogously than as in the proof of Lemma 24)

$$\Leftrightarrow a \vee x \vee g(x) = a \vee x \vee h(x) \Leftrightarrow a \vee g(x) = a \vee h(x).$$

Since $D \in R$ (A) we deduce that g(x) =h(x) for every $X \in$ dom $\cap$ dom (h) so [g] = [h], which is contradictory. Hence, if [g] $\neq$ [h], then there exists $a \in D \cap B$ (A), such that $\bar{f_a} \vee [g] \neq \bar{f_a} \vee [h]$. But for this a $a \in D \cap B(A)$ we have $\bar{f_a} \vee [f])(x) \in B(A)$ (since $D \cap B(A) \subseteq D_{[f]}$) hence $A \succ Q$ (A). To prove the maximality of Q(A), let A' be a bounded commutative BCK-algebra such that $A \succ A'$, thus $B(A) \subseteq B(A')$; Then A' is embedded in Q(A) by $i : A' \rightarrow Q(A)$ defined by $i : (a') = [f'_a]$, for every $a' \in A'$, where dom$(f'_a) \in D_a$. (See Lemma 28).

Clearly, $f'_a \in M_r(A)$ (by Lemma 28) and i is a morphism of BCK-algebras (see Lemma 18). To prove the injectivity of i, let $a', b' \in A'$ such that $[f_{a'}] = [f_{b'}] \Leftrightarrow f_{a'}(x) = f_{b'}(x)$ for every $x \in D_{a'} \cap D_{b'}$. If $a' \neq b'$ by Lemma 27 (since $A \succ A'$), there exists $a \in B(A)$ such that $a \vee a', a \vee b' \in B(A)$ and $a \vee a' \neq a \vee b'$, which is contradictory (since $a \vee a', a \vee b' \in B(A)$ implies $a \in D_{a'} \cap D_{b',}$).

Remark 10:

1. If A is a BCK-algebra with B (A) = {0, 1} =$L_2$ and $A \succ A'$ then A'= {0, 1}, hence Q (A) $\approx L_2$. Indeed, if a, b, c$\in$ A', with a$\neq$b, then there exists e$\in$B (A) such that $e \vee a \neq e \vee b$, (hence e$\neq$ 1) and$e \vee c \in B(A)$. Clearly, e=0 hence c$\in$B (A), that is A' =B (A). As examples of BCK-algebras with this property we have local BCK-algebras and BCK-chains.

2. More general, if A is a BCK-algebra such that B(A) is finite, if $A \succ A'$ then $A' = B(A)$ hence $Q(A) = B(A)$

Indeed, $B(A) \subseteq A'$ and consider $a \in A'$. $B(A)$ being finite, there exists a smallest element $e_a \in B(A)$ such $e_a \vee a \in B(A)$. Suppose $e_a \vee a \neq a$, then there would exists $e \in B(A)$ such that $e \vee (e_a \vee a) \neq e \vee a$ and $e \vee a \in B(A)$. But $e \vee a \in B(A)$ implies $e_a \leq e$, and thus we obtain $e \vee (e_a \vee a) \neq e \vee a \Leftrightarrow e \vee a \neq e \vee a$. a contradithon. Hence $e = e_a \vee a \in B(A)$, that is, $A' \subseteq B$ (A). Then $A' = B$ (A), hence $Q$ (A)= B(A).

## LOCALIZATION OF COMMUTATIVE BOUNDED BCK-ALGEBRAS

In [4], G. Georgescu exhibited the localization lattice $L_F$ of a distributive lattice L with respect to a topology F on L in a similar way as for rings (see [16]) or monoids (see [17]). The aim of this section is to define the notion of localization BCK-algebra $A_F$ of a commutative bounded BCK-algebra A with respect to a topology F on A. In the last part of this section is proved that the maximal commutative BCK-algebra of quotients (defined in Section 5) and the commutative BCK-algebra of fractions relative to a V -closed system (defined in Section 3) are BCK-algebras of localization.

In this section A will be a bounded commutative BCK-algebra and F a topological system on A.

**Definition 12:** A non-empty family F of elements on T(A) will be called a topological system on A if the following properties hold:

$(t_1)$ If $D_1 \in F, D_2 \in T(A)$ and $D_1 \subseteq D_2$, then $D_2 \in F$ (hence $A \in F$);

$(t_2)$ If $D_1, D_2 \in F$, then $D_1 \subseteq D_2 \in F$.

Example 3: If $D \in T(A)$, then the set $F_D = \{D' \in T(A) : D \subseteq D\}$ is a topological system on A.

Example 4: We recall that by R (A) we denote the set of all regular subsets of A (see Definition 8). Then $F = T(A) \cap R(A)$ is a topological system on A (see Lemma 19).

Example 5: Let $S \subseteq A$ a V-closed subset of A (see Definition 4). If we denote by $F_S = \{D \in T(A) : D \cap S \cap B(A) \neq \emptyset\}$, then $F_s$ is a topological system on A.

If F is a topological system on A, let us consider the relation $\theta_F$ of A defined by: $(x, y) \in \theta_F \Leftrightarrow$ there exists $D \in F$ such that $t \vee x = t \vee y$ for any $t \in D \cap B(A)$. As in the case of $\theta_S$ (see Proposition 11), we deduce that $\theta_F$ is a congruence on A.

We shall denote by $x/\theta_F$ the congruence class of an element $x \in A$ and by $p_F : A \rightarrow A/\theta_F$ the canonical morphism of BCK-algebras.

Remark 11: Clearly, if $a \in B(A) \Rightarrow a/\theta_F \in B(A/\theta_F)$.

**Definition 13:** A F -multiplier on A is a mapping $f : D \rightarrow A/\theta_F$ where $D \in F$ such that for every $a \in B(A)$ and $x \in D$,

$(m_1) =). f(a \vee x) = a/\theta_F \vee f(x);$

$(m_2) \ x/\theta_F \leq f(x).$

If F={A}, then a F-multiplier is a function $f: A \to A$ which verify only the conditions $sa_1$ and $sm_2$ from Definition 7. The maps 0, 1: $A \to A / \theta_F$, defined by $0(x) = x / \theta_F$ and $1(x) = 1 / \theta_F$ for every $x \in A$ are F-multipliers. Also, for $a \in B(A)$, $f_a : D \to A / \theta_F$ defined by $f_a(x) = a / \theta_F \vee x / \theta_F$ for every $x \in D$, is a F-multiplier (where $D \in F$).

For $D \in F$, we shall denote by $M(D, A / \theta_F)$ the set of all the F-multipliers having the domain D. If $D_1$, $D_2 \theta_F F$, $D_1 \subseteq D_2$ we have a canonical mapping $\phi_{D_1,D_2} : M(D_2, A / \theta_F) \to M(D_1, A / \theta_F)$ defined by $\phi_{D_1,D_2}(f) : f_{|D_1}$ for $f \in M(D_2, A / \theta_F)$. Let us consider the directed system of sets $< \{M(D, A / \theta_F)\}_{D \in F}, \{\phi_{D_1,D_2}\}_{D_1,D_2 \in FD_1 \subseteq D_2} >$ and denote by $A_F$ the inductive limit Cm the category of sets): $A_F = \overset{\lim}{\underset{D \in F}{\longrightarrow}} M(D, A / \theta_F)$ For any F-multiplier $f : D \to A / \theta_F$ we shall denote by $\overline{(D,f)}$ the equivalence class of f in $A_F$.

Remark 12: We recall that if $f_i : D_i \to A / \theta_F, i = 1,2$ are F-multipliers, then $\overline{(D_1,f_1)} = \overline{(D_2,f_2)}$ (in $A_F$) if there exists $D \in F, D \subseteq D_1 \cap D_2$ such that. $f_{1|D} = f_{2|D}$

Let: $F_i : D_i \to A / \theta_F$, (with $D_i \in F$, i= 1,2), F-multipliers. Let us consider the mapping $f_1 \to f_2 : D_1 \cap D_2 \to A / \theta_F$, defined by $(f_1 \to f_2)(x) = f_1(x) \to f_1(x)$, for any $x \in D_1 \cap D_2$, and let

$$\overline{(D_1,f_1)} \to \overline{(D_2,f_2)} = \overline{(D_1 \cap D_2, f_1 \to f_2)}.$$

This definition is correct. Indeed, let $f'_i : D'_i \rightarrow A / \theta_F$, with $D'_i \in F$, i $=1,2$ such that $\overline{(D_i, f_i)} = \overline{(D'_i, f'_i)}$, i $=1,2$. Then there exist $D''_1, D''_2 \in F$ such that $D''_1 \subseteq D_1 \cap D'_1, D''_2 \subseteq D_2 \cap D'_2$; and $f_{1|D''_1} = f'_{1|D''_1}, f_{2|D''_2} = f'_{2|D''_2}$. If we set $D'' = D''_1 \cap D''_2 \subseteq D_1 \cap D_2 \cap D'_1 \cap D'_2$, then $D'' \in F$ and clearly $(f_1 \rightarrow f_2)_{|D''} = (f'_1 \rightarrow f'_2)_{|D''}$.

Hence

$$\overline{(D_1 \cap D_2, f_1 \rightarrow f_2)} = \overline{(D'_1 \cap D'_2, f'_1 \rightarrow f'2)}.$$

Lemma 30: $f_1 \rightarrow f_2 \in M(D_1 \cap D_2, A/\theta_F)$.

Proof: If $x \in D_1 \cap D_2$ and $a \in B(A)$, then

$$(f_1 \rightarrow f_2)(a \vee x) = f_1(a \vee x) \rightarrow f_2(a \vee x)$$
$$= (a/\theta_F / \vee f_1(x)) \rightarrow (a/\theta_F \vee f_2(x))$$
$$\overset{(c_{21})}{=} a/\theta_F \vee (f_1 \rightarrow f_2)(x)$$

and

$$(f_1 \rightarrow f_2)(x) = f_1(x) \rightarrow f_2(x) \geq f_2(x) \overset{(m_2)}{\geq} x/\theta_F.$$

Corollary 31: $(A_F \rightarrow, \overline{0}, \overline{1})$ is a bounded commutative BCK-algebra (where $\overline{0} = \overline{(A, 0)}$ and $\overline{1} = \overline{(A, 1)})$ (see Corollary 17).

**Definition 14:** $A_F$ will be called the localization BCK-algebra of A with respect to the topology F.

Lemma 32: The mapping $v_F : B(A) \rightarrow A_F$ defined by $v_F(a) = \overline{(A, \overline{f_a})}$ for every $a \in B$ (A) is a morphism of BCK-algebras and $v_F(B(A))$ is a regular subset of $A_F$.

Proof: If a, $b \in B$ (A) then

$$v_F(a) \rightarrow v_F(b) = \overline{\left( A, \overline{f_a} \right)} \rightarrow \overline{\left( A, \overline{f_b} \right)}$$

$$= \overline{\left( A, \overline{f_a} \rightarrow \overline{f_b} \right)} = \overline{\left( A, \overline{f_{a \rightarrow b}} \right)} = v_F(a \rightarrow b).$$

To prove that $v_F$ (A) is a regular subset of $A_F$, let $(D_i, f_i) \in A_F, D_i \in F, i = 1, 2$, such that

$$\overline{\left( A, \overline{f_a} \right)} \vee \overline{(D_1, f_1)} = \overline{\left( A, \overline{f_a} \right)} \vee \overline{(D_2, f_2)}$$

for every $a \in B(A)$.

Then $\overline{(D_1, f_a \vee f_1)} = \overline{(D_2, f_a \vee f_2)} \Leftrightarrow$ there exists $D \in F, D \subseteq D_1 \cap D_2$ such that

$$\left( f_a \vee f_1 \right)_{|D} = \left( f_a \vee f_2 \right)_{|D}$$

$$\Leftrightarrow (a \vee x)/\theta_F \vee f_1(x) = (a \vee x)/\theta_F \vee f_2(x),$$

for every $x \in D$ and $a \in B(A)$. If in this last equivalence we choose a = $0 \in B(A)$, then we obtain that

$$x/\theta_F \vee f_1(x) = x/\theta_F \vee f_2(x)$$
$$\Leftrightarrow f_1(x) = f_2(x) \Leftrightarrow \overline{(D_1, f_1)} = \overline{(D_2, f_2)},$$

hence $v_F(B(A))$ is a regular subset of $A_F$.

## APPLICATIONS

In that follows we describe the localization BCK-algebra $A_F$ in some special instances.

1. If $D \in T(A)$ and $F$ is the topological system $F_D = \{D' \in T(A) : D \subseteq D'\}$ (see Example 3), then $A_F \subseteq M(D, A/\theta_F)$ and $v_F : B(A) \to A_F$ is defined by $v_F(a) = (D, f_{a|D})$ for any $a \in B(A)$. For x, y $\in$ B(A) we have $(x, y) \in \theta_F \Leftrightarrow$ for every $t \in D, t \vee x = t \vee y \Leftrightarrow f_{x|D} = f_{y|D} \Leftrightarrow v_F(x) = v_F(y)$

   then there exists an injective morphism of BCK-algebras $\varphi : A/\theta_F \to A_F$, $\varphi(x/\theta_F) = v_F(x)$ such that $\varphi \circ v_F = p_F$.

2. To obtain the maximal BCK-algebra of quotients Q (A) as a localization relative to a topological system F we will develop another theory of F-multipliers (meaning we add new axioms for F-multipliers).

**Definition 15:** Let F be a topological system on A. A strong-F-multiplier is a mapping $f : D \to A/\theta_F$ (where D $\in$ F) which verifies the axioms $m_1$ and $m_2$ and $(m_3)$ If $e \in D \cap B(A)$, , then $f(e) \in B(A / \theta_F); (m_4)(x / \theta_F) \vee f(a) = (a / \theta_F) \vee f(x)$, , for every $a \in D \cap B(A)$ and $x \in D$.

If $F = \{A\}$, then $\theta_F$ is the identity congruence of A so a strong F-multiplier is a strong total multiplier (in sense of Definition 7).

Remark 13: If A is a BCK-algebra, the maps 0, 1: $A \to A / \theta_F$ defined by 0 $(x) = x / \theta_F$ and 1 $(x) = 1 / \theta_F$ for every $x \in A$ are strong F-multipliers. If $f_i : D_i \to A / \theta_F$, (with $D_i \in F, i = 1, 2$) are strong F-multipliers, the mapping $f_1 \to f_2 : D_1 \cap D_2 \to A / \theta_F$ defined by $(f_1 \to f_2)(x) = f_1(x) \to f_2(x)$, for any $x \in D_1 \cap D_2$ is also a strong-F-multiplier.

Remark 14: Analogous as in the case of F-multipliers if we work with strong-F-multipliers we obtain a BCK-subalgebra of $A_F$ denoted by s $— A_F$ which will be called the strong localization BCK-algebra of A with respect to the topological system F.

If $F = T (A) \cap R (A)$, then $\theta_F$ is the identity congruence of A and we

obtain the definition for strong multipliers on A, so $A_F = \lim M(D, A)$. In this situation it is easy to see that $v_F$ is injective, so we have:

Proposition 33: In the case $F = I(A) \cap R(A), s - A_F$ is exactly the maximal commutative BCK-algebra of quotients Q (A) of A (see Section 5, Theorem 29).

3. Let S be a V-closed system of A. We recall (see Proposition 11) that on A we have the congruence $\theta_s$ defined by: $(x, y) \in \theta_s$ iff there is $s \in S \cap B(A)$ such that $s \vee x = s \vee y$ and A [S] $= A / \theta_s$ is called the (commutative) BCK-

algebra of fractions of A relative to the V -closed system S (see Remark 5 from Section 3). In this case we have the topological system $F_s$ associated with S, $F_s = \{D \in T(A) : D \cap S \cap B(A) \neq \varnothing\}$.

Lemma 34: $\theta_{F_s} = \theta_S$.

Proof: For x, y $\in$ A, if (x, y) $\in \theta_{F_s}$ then there exists $D \in F_s$ such that $s \vee x = s \vee y$ for every $s \in S \cap B(A)$.

Since $D \in F_s, D \cap S \cap B(A) \neq \varnothing$, so there exists $s_0 \in D \cap S \cap B(A)$; in particular we obtain $s_0 \vee x = s_0 \vee y$, hence (x, y) $\in \theta_S$, that is, $\theta_{F_s} \subseteq \theta_S$.

If (x, y) $\in \theta_S$, then so $s_0 \vee x = s_0 \vee y$, for some $s_0 \in S \cap B(A)$. If consider $D = [s_0 = \{a \in A : s_0 \leq a\}]$ (the principal deductive system generated by $s_0$, see Corollary 8, 1)), then $D \in F_s$ (since $s_0 \in D \cap S \cap B(A)$). If $s \in S \cap B(A)$ then $s_0 \leq s \Rightarrow s = s \vee s_0$ hence

$$s \vee x = \left(s \vee s_0\right) \vee x = s \vee \left(s_0 \vee x\right)$$
$$= s \vee \left(s_0 \vee y\right) = \left(s \vee s_0\right) \vee y = s \vee y$$
$$\Rightarrow \left(x, y\right) \in \theta_{F_s} \Rightarrow \theta_S \subseteq \theta_{F_s} \Rightarrow \theta_{F_s} = \theta_S.$$

Proposition 35: If Fs is the topological system on A associated with a V -closed subset S of A, then s — $A_{F_s}$ is isomorphic with B (A[S]).

Proof: Following Lemma 34, $\theta_{F_s} = \theta_S$, therefore a $F_s$-multiplier can be considered in this case as a mapping $f : D \to A[S]$ $(D \in F_s)$ having for x $\in$ D and a $\in D \cap B(A)$ the properties

$$f(a \vee x) = a/\theta_S \vee f(x)$$
$$= x/\theta_S \vee f(a), x/\theta_S \leq f(x),$$
$$f(a) \in B(A[S]).$$

If

$$\overline{(D_1, f_1)}, \overline{(D_2, f_2)} \in s - A_{F_s} = \varinjlim_{D \in F_s} M(D, A[S]),$$

and $\overline{(D_1, f_1)} = \overline{(D_2, f_2)}$ then there exists $D \in F_s$ such that $D \subseteq D_1 \cap D_2$ and $f_{1|D} = f_{2|D}$. Since D, $D_1$, $D_2 \in F_s$, then

$$D \cap S \cap B(A), D_1 \cap S \cap B(A), D_2 \cap S \cap B(A)$$

are nonempty, hence there exist $s \in D \cap S \cap B(A)$, $s_1 \in D_1 \cap S \cap B(A)$ and $s_2 \in D_2 \cap S \cap B(A)$.

We shall prove that $f_1(s_1) = f_2(s_2)$. Indeed, if consider

$$t = s \vee s_1 \vee s_2 \in D \cap S \cap B(A)$$

then

$$f_1(t) = s/\theta_S \vee s_2/\theta_S \vee f_1(s_1) = f_1(s_1)$$

(since $s/\theta_S = s_2/\theta_S = 0$) and analogously $f_2(t) = f_2(s_2) \Rightarrow f_1(s_1) = f_2(s_2)$. In a similar way we can show that $f_1(t_1) = f_2(t_2)$ for any $t_1, t_2 \in D \cap S \cap B(A)$. In accordance with these considerations we can define the mapping

$$\alpha : s - A_{F_S} = \varprojlim_{D \in F_S} M(D, A[S]) \to B(A[S])$$

by putting

$$\alpha\overline{(D, f)} = f(s),$$

where $s \in D \cap S \cap B(A)$. It is easy to prove that $\alpha$ is a morphism of BCK-algebras. We shall prove that $\alpha$ is injective and surjective. To prove the injectivity of $\alpha$ let $\overline{(D_1, f_1)}, \overline{(D_2, f_2)} \in s - A_{F_S}$ such that $\alpha(\overline{(D_1, f_1)}) = \alpha(\overline{(D_2, f_2)})$. Then for any $S_1 \in D_1 \cap S \cap B(A), S_2 \in D_2 \cap S \cap B(A)$ we have $f_1(S_1) = f_2(S_2)$. For two fried elements $s_1, s_2$, with $S_i \in D_i \cap S \cap B(A), i = 1, 2$, we consider the element $s = s_1 \vee s_2 \in (D_1 \cap D_2) \cap S \cap B(A)$. We have

$$f_1(s) = s_2/\theta_S \vee f_1(s_1) = 0 \vee f_1(s_1) = f_1(s_1)$$

and

$$f_2(s) = s_1/\theta_S \vee f_2(s_2) = 0 \vee f_2(s_2) = f_2(s_2),$$

hence $f_1(s) = f_2(s)$. Now let

$$D_s = [s) \cap D_1 \cap D_2 = \{s' \in D_1 \cap D_2 : s \le s'\}.$$

Since $s \in D$, we deduce that $D_s \ne \varnothing$. If $a \in A$ and $s' \in D_s$, then $s \le s' \le a \vee s' \Rightarrow a \vee s' \in D_s$. Since $s \in D_s \cap D \Rightarrow D_s \in F_s$. If $s' \in D_s$ then

$$s \vee s' = s' \Rightarrow f_1(s') = f_1(s \vee s')$$
$$= s'/\theta_s \vee f_1(s) = 0 \vee f_1(s) = f_1(s)$$

and analogously,

$$f_2(s') = f_2(s) \Rightarrow f_1(s') = f_2(s')$$
$$\Rightarrow f_{1|Ds} = f_{2|Ds} \Rightarrow \overline{(D_1, f_1)} = \overline{(D_2, f_2)},$$

that is, $\alpha$ is injective. To prove the siujectivity of $\alpha$, let $a / \theta_s \in B(A[S])$ with $a \in A$.

For one fixed element $s \in S$, we consider $D = [s) = \{x \in A : s \le x\}$. Clearly $D \in F_s$. We define $f_a : D \to A[S]$ by putting $f_a(x) = (a \vee x) / \theta_F$, for every $x \in D$. Clearly, $f_a$, is a strong $F_s$-multiplier (clearly $(m_3)$ is verified since if $e \in D \cap B$ (A), then $f_a(a) = (a \vee e) / \theta_s = a / \theta_s \vee e / \theta_s \in B(A[S])$. From

$$\left(a \vee s\right) \vee s = a \vee s \Rightarrow \left(a \vee s\right)/\theta_S = a/\theta_S$$

$$\Rightarrow f_a\left(s\right) = a/\theta_S \Rightarrow \alpha\overline{\left(\left(D, f_a\right)\right)} = a/\theta_S,$$

that is, $\alpha$ is surjective, hence bijective.

## REFERENCES

1. D. Piciu and C. Busneag, "The Localization of Commutative (Unbounded) Hilbert Algebras," Mathematical Reports, Vol. 12, No. 3, 2010, pp. 285-300.

2. Y. Imai and K. Iséki, "On Axiom Systems of Propositional Calculi XIV," Proceedings of the Japan Academy, Vol. 42, No. 1, 1966, pp. 19-22. doi:10.3792/pja/1195522169

3. N. Prior, "Formal logic, 2nd Ed," Oxford, 1962.

4. G. Georgescu, "F-Multipliers and the Localization of Distributive Lattices," Algebra Universalis, Vol. 21, No. 2-3, 1985, pp. 181-197. doi:10.1007/BF01188055

5. S. Rudeanu, "Localizations and Fractions in Algebra of Logic," Journal of Multi-Valued Logic & Soft Computing, Vol. 16, No. 3-5, 2010, pp. 465-467.

6. R. Cignoli and A. Torens, "Glivenko Like Theorems in Natural Expansions of BCK-Logic," Mathematical Logic Quarterly, Vol. 50, No. 2, 2004, pp. 111-125. doi:10.1002/malq.200310082

7. J. Gispert and A. Torrens, "Boolean Representation of Bounded BCK-Algebras," Soft Computing, Vol. 12, No. 10, 2008, pp. 941-954. doi:10.1007/s00500-007-0261-0

8. Iorgulescu, "Algebras of Logic as BCK-Algebras," ASE, Bucharest, 2008.

9. K. Iséki and S. Tanaka, "An Introduction to the Theory of BCK-Algebras," Mathematica Japonica, Vol. 23, No. 1, 1978, pp. 1-25.

10. F. M. G. Olmedo and A. J. R. Salas, "Negation and BCK- Algebras," Mathematical Logic Quarterly, Vol. 49, No. 4, 2003, pp. 336-346. doi:10.1002/malq.200310035

11. J. Schmid, "Multipliers on Distributive Lattices and Rings of Quotients," Houston Journal of Mathematics, Vol. 6, No. 3, 1980, pp. 401-425.

12. J. Schmid, "Distributive Lattices and Rings of Quotients," Colloquia Mathematica Societatis János Bolyai, Vol. 33, Szeged, Hungary, 1980.

13. J. Lambek, "Lectures on Rings and Modules," Blaisdell Publishing Company, New York, 1966.

14. H. Cornish, "The Multiplier Extension of a Distributive Lattice," Journal of Algebra, Vol. 32, No. 2, 1974, pp. 339-355. doi:10.1016/0021-8693(74)90143-4

15. R. Balbes and Ph. Dwinger, "Distributive Lattices," University of Missouri Press, Columbia, 1974.

16. N. Popescu, "Abelian Categories with Applications to Rings and Modules," Academic Press, New York, 1973.

17. B. Strenstr?m, "Platnes and Localization over Monoids," Mathematische Nachrichten, Vol. 48, No. 1-6, 1971, pp. 315-334. doi:10.1002/mana.19710480124

## CITATION

D. Piciu and D. Tascau, "The Localization of Commutative Bounded BCK-Algebras," Advances in Pure Mathematics, Vol. 1 No. 6, 2011, pp. 367-377. doi: 10.4236/apm.2011.16066.

# A Linear Interpolation-Based Algorithm for Path Planning and Replanning on Girds

*Changwen Zheng, Jiawei Cai, and Huafei Yin*

National Key Laboratory of Integrated Information System Technology Institute of Software, Chinese Academy of Sciences, Beijing, China

## ABSTRACT

Field D* algorithm is widely used in mobile robot navigation since it can plan and replan any-angle paths through nonuniform cost grids. However, it still suffers from inefficiency and sub-optimality. In this article, a new linear interpolation-based planning and replanning algorithm, Update-Reducing Field D*, is proposed. It employs different approaches during initial planning and replanning respectively in order to reduce the number of updates of the rhs-values of vertices. Experiments have shown that Update-Reducing Field D* runs faster than Field D* and returns smoother and lower-cost paths.

## INTRODUCTION

In mobile robot navigation, path planning leads a robot from its initial location to some desired goal location. The two most popular techniques for path planning are deterministic algorithms and randomized algorithms [1]. Among deterministic algorithms, A* provides heuristic search in static, known environments [2].LPA* combines heuristic search and incremental search [3]. D* Lite could replan in unknown environments efficiently [4].

When provided with a grid-based representation of environments, these algorithms are limited by the discrete set of possible headings between gird cells. For example, the eight-connected grids restrict the agent's heading changes by multiples of $\pi / 4$. As a result, the paths are suboptimal and unrealistic looking. To alleviate this problem, several methods for any-angle path planning have been investigated. A* PS uses post-smoothing to generate any-angle path [5]. But it does not always work as is showed in Figure 1 (The path returned by A* PS (in black line) is not the optimal path (in dash line) from S to G). Basic Theta* algorithm and AP Theta* algorithm (Angle-Propagation Theta*) allow the parent of a node to be a node other than its local neighbor [6]. AA* (Accelerated A*) can plan a shortest any angle paths fast [7]. However, these algorithm are only usable for uniform cost environments.

Based on D* Lite and linear interpolation, Field D* could fast plan and re-plan any-angle path in grid environments, whatever the environment costs are known or partially-known, uniform or non-uniform [8]. Field D* is employed as the path replanner in a wide range of fielded robotic systems.

However, Field D* still suffers from two major drawbacks: 1) It plans and replans much slower than D* Lite. 2) The path returned by Field D* is not always the optimal solution. Motivated by these observations, a linear interpolation-based planning and replanning algorithm, Update-Reducing Field D* (URFD*), is proposed in this paper. It reduces the number of updates of the rhs-values so as to speed up the search. It also employs a method of post-smoothing to generate a lower-cost and smoother path. Besides, a heuristic with a variable factor according to the environments is used. As a result, the novel algorithm could efficiently produce a near-optimal path in non-uniform cost grids.

## LINEAR INTERPOLATION-BASED PATH PLANNING

### The Idea of Field D*

Field D* stores the rhs-value, a one-step look ahead estimate of the goal distance (by the goal distance of a vertex we mean the cost of an optimal path from this vertex to the goal). For vertex s it satisfies:

$$rhs(s) = \begin{cases} 0, & \text{if } s = s_{goal} \\ \min_{s' \in nbrs(s)} \left( g(s') + c(s',s) \right) & \text{otherwise} \end{cases}$$

(1)

Where $s_{goal}$ is the goal vertex. Nbrs(s) Denotes the set of all neighboring vertices of s. $g(s')$ is an estimate of goal distance of s'. $c(s',s)$ Is the cost of a path between s' and s. In classical grid-based methods, it is assumed that s' and s are two corner vertices and the path between s' and s is a straight line. Field D* relaxes this assumption and takes any point along the boundary of a cell into consideration. To make it possible, Field D* makes an approximation that the path cost of any point $s_y$ residing on the edge between two consecutive corner vertices $s_1$ and $s_2$ is a linear combination of $g(s_1)$ and $g(s_2)$:

$$g(s_y) = yg(s_2) + (1-y)g(s_1)$$

(2)

Where y is the distance from $s_1$ to $s_y$ (assuming unit cells).

With the form of the optimal path in a unit cell, Field D* could compute rhs(s) and find the point to move by making

$$d(c(s',s) + g(s'))/dv = 0,$$

(3)

Where v is the variable on which the path cost depends.

## Differences and Inefficiency

A linear interpolation-based replanner performs differently from a classical path replanner such as D* Lite, which leads to its inefficiency. To explain this, we define $g^*(s)$ as the cost of the optimal path from vertex s to the goal with respect to the linear interpolation assumption (so it is slightly different from the cost of the actual optimal path). We call

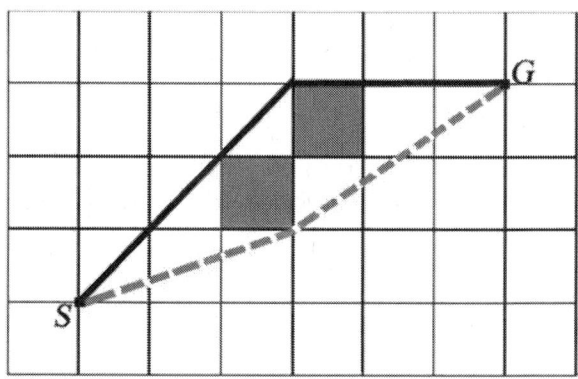

**Figure 1**: Sub-Optimality of A* PS.

g(s) (or rhs(s)) is inaccurate when g(s) (or rhs(s)) is not equal to $g^*(s)$. Also we call g(s) is more accurate than rhs(s) when g(s) is more close to $g^*(s)$ than rhs(s). $g^*(s)$ Satisfies:

$$g^*(s) = \begin{cases} 0, & \text{if } s = s_{goal} \\ \min_{s' \in nbrs(s)} \left( g^*(s') + c(s', s) \right) & \text{otherwise.} \end{cases}$$

(4)

A linear interpolation-based replanner expands vertices in a different way from D* Lite. With a consistent heuristic, a locally overconsistent vertex (whose g-value is lager than rhs-value) becomes locally consistent (the g-value equals rhs-value) after selected for expansion and then remains locally consistent until edge cost changes are detected in D* Lite. It implies that D* Lite expands any locally overconsistent vertex at most once. However, a linear interpolation-based replanner tends to expand a locally overconsistent vertex for many times. Besides, the key values (denoting the priorities of vertices in the priority queue) of the vertices selected for expansion are monotonically non-decreasing over time in D* Lite, while it is not naturally the case in a linear interpolation-based replanner. This can be explained as follows: For some vertex s, the computation of its rhs-value is based on the g-value of one neighboring vertex in D* Lite, but the g-values of two neighboring vertices in a linear interpolation-based replanner. During the planning and replanning process, it is common that at least one of the two neighboring vertices has an inaccurate rhs-value. Relied upon them, s will get an inaccurate rhs-value. And the vertices relied upon s will also be affected, and so on. It is the reason that a linear interpolation-based replanner updates the rhs-values of vertices repeatedly to their final results. Such a phenomenon is easily observed particularly in environment consisting mainly of free space since the g-value of a vertex is very close to those of its neighbors in free space.

After the cost field created, path extraction for linear interpolation-based replanners is also different from that for D* Lite. When backpointers are not recorded, D* Lite can trace back a lowest-cost path from $s_{start}$ to $s_{goal}$ by always moving from the current vertex s, starting at $s_{start}$, to any neighbor s' that minimizes $c(s',s)+g(s')$ until $s_{goal}$ is reached. So the optimality of the result depends on the accuracy of $g(s')$. But in a linear interpolationbased replanner the g-values are not the goal distances exactly, resulting in the sub-optimality of paths.

From the discussion above, we can see that there exist two major drawbacks of Field D*: 1) It plans and replans much slower than D* Lite, especially in the environments consisting mainly of free space. 2) The solution path is not the optimal solution. Figure 2 shows the second problem. The path returned by Field D* has unnecessary heading changes even if no obstacle exists (see Figure 2(a)). To extract a smoother path, [9] gives a gradient interpolation method. The result is showed in Figure 2(b), from which we can see that the unnecessary heading changes still exit. Obstacles can also make the interpolation assumption break down so that affects the quality of extracted paths. To alleviate it, [8] uses an onestep look ahead mechanism.But this method checks very limited steps so that cannot avoid generating a pathologic path between vertices a and b in Figure 2(c).

## UPDATE-REDUCING FIELD D*

### Basic Idea

Since the runtime of a linear interpolation-based replanner depends heavily on the number of updates of the rhsvalues of vertices, the key to high efficiency of our algorithm is reducing the number of updates. We use the fringe vertices to refer to the vertices on the fringe of the expanded vertices during initial planning. The fringe vertices have not been expanded yet so that their g-values are not computed (namely infinite), leading the rhs-values computed by the g-values of the fringe vertices to inaccuracy. Then the inaccurate rhs-values along with the infinite g-values affect next vertex expansions. This is the main source of the repeated updates of rhs-values. Note that most of the fringe vertices are just locally overconsistent vertices during initial planning, as is showed in Figure 3. (The locally consistent vertices (in light grey), which have been expanded at least once, are almost surrounded by the locally overconsistent vertices (in dark grey). Vertices in black are obstacles.) The rhs-values of locally overconsistent vertices are better informed and thus more accurate than the g-values. So when it is a locally overconsistent vertex, we can use the rhs-value instead of the g-value to make the computation more close to the g*-value. When it is a locally consistent vertex, we can also use the rhs-value because it equals the g-value and thus could get a result at least no poorer than that computed by the g-value.

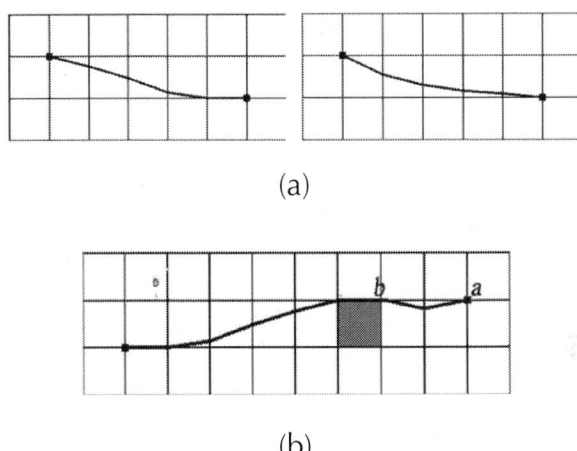

(a)

(b)

**Figure 2**: Paths returned by interpolation methods.

During replanning, if we only encounter cell cost decreases the approach above is still useful. However, when locally underconsistent vertices (whose g-values are smaller than rhs-values) appear, this approach tends to make the algorithm less efficient and even incomplete. It could be explained as follows: When the rhs-value of a vertex becomes larger due to edge cost increases, the old rhs-value is out of date and thus to be abandoned. However, the algorithm does not distinguish between the old and the new so that it is possible for the old rhs-value to be used to compute the rhs-value of another vertex, resulting in "false" relation between these two vertices. For example, there exist two vertices a and b. After initial planning, g (a), rhs(a) and g(b) are all infinite, rhs(b) is 50. Then rhs(a) and rhs(b) are updated due to cell cost increases. When rhs(a) is recomputed, old rhs(b) (namely 50) is used. Then rhs(b) is updated to a new value of 64. Thus, the computation of rhs(a) seems to rely on rhs(b) but in fact this relation possibly dose not exist. Furthermore the wrong rhs(a) leads to a priority error of a (g(a) is infinite so that does not affect the key value), which possibly makes a expand while b can never be expanded again. Thus the "false" relation has no chance to be corrected, resulting in the incompleteness of the algorithm.

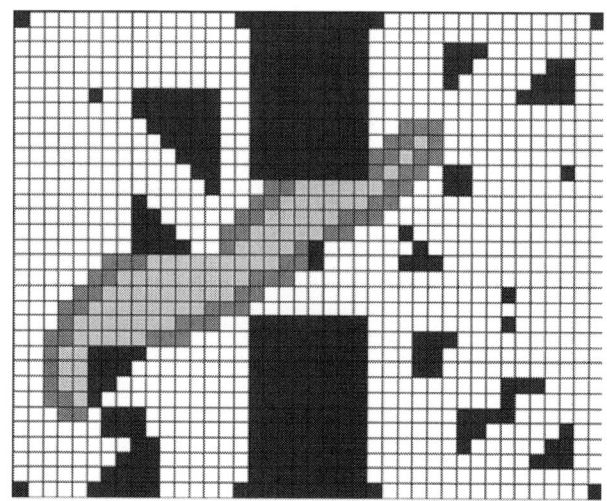

**Figure 3**: A snapshot during initial planning of Field D*.

In order to reduce the updates of the rhs-values during replanning, we use a technique similar to which Delayed D* used to speed up D* Lite [10], that is, delaying the processing of locally underconsistent vertices.

During path extraction, A* search, which depends on the accuracy of heuristic less heavily than greedy search does, could avoid errors caused by obstacles (see Figure 2(c)). However, based on the linear approximation, A* search still cannot ensure an optimal path even if no obstacles exist (see Figure 2(b)). And greedy search needs to be kept for checking solution paths for any loops. Note that the limitation of post-smoothing showed in Figure 1 can be overcome if it is already an any-angle path before smoothing.

Combined with the methods above, Update-Reducing Field D* (URFD*) is a modified version of Field D*. It redefines the rhs-values (denoted by rhs'-values to be distinguished from the original) as

$$rhs'(s) = \begin{cases} 0 & \text{if } s = s_{goal} \\ \min_{s' \in nbrs(s)} \left( rhs'(s') + c(s',s) \right) & \text{otherwise} \end{cases}$$

(5)

Where notation follows from (1). URFD* calculates the rhs-values of vertices according to (5) during initial planning. During replanning it calculates the rhs-values according to (1), which is similar to Field D*, and delays the propagation of cost increases. It checks the consistency of a path in every path extraction and ends with a post-smoothing step.

## Algorithm Description

**Figure 4** shows the pseudocode of the URFD* algorithm. During initial planning, URFD* calls ComputeShortestPath() to expand vertices. ComputeCost() calculates the rhs-value in a way similar to the interpolation-based path cost calculation in Field D*, but every vertex uses rhsvalues, instead of g-values, of its neighbors. ComputeState() then computes the rhs-values according to (5) (line 15). During replanning, ComputeState() calls ComputeCost() to compute the rhs-values according to (1) (line 16). The rhs-values of the start vertex and every vertex immediately affected by the changed edge costs are updated, but only the locally inconsistent start vertex and locally overconsistent vertices are inserted into priority queue U for expansion (lines 42 - 45). Then FindRaiseStatesOnPath() (FRSOP) is called. FROSP checks whether locally underconsistent vertices are in the vicinity of the node. All the unprocessed locally underconsistent vertices that are adjacent to this node will be added into priority queue U (lines 31 - 33). Here vicinity(s) refers to the set of all corner vertices in the vicinity of node s (s is included). When the number of nodes exceeds the given limit maxsteps, or a loop is found, which indicates a potential failure of path extraction, FRSOP stops the extraction to expand locally underconsistent vertices in priority queue U. After path extraction, the solution path is post-processed by a smoothing step. (Lines 39, 49). Given two cell boundary nodes along the path, the postsmoothing replaces the solution path between these two nodes with a straight line path if the latter is less costly. It is done with cell boundary nodes along the solution path iteratively. However, some small techniques are used to avoid a large amount of computation: 1) It only performs a single iteration. 2) It only smoothes the path between cell corners because necessary and sharp heading changes usually occur on them. 3) Before smoothing a

**Key(s)**
01. return $[\min(g(s), rhs(s)) + h(s_{start}, s); \min(g(s), rhs(s))]$;

**Initialize(s)**
02. for all $s \in S$ $rhs(s) = g(s) = \infty$;
03. $rhs(s_{goal}) = 0$; $U = \varnothing$; $raise = false$;
04. Insert$(U, s_{goal}, \text{Key}(s_{goal}))$;

**ComputeCost$(s, s_a, s_b, cost)$**
05. Use $cost(s_a)$ and $cost(s_b)$ to compute $rhs(s)$;

**UpdateState(s)**
06. if$(g(s) \neq rhs(s))$
07.   if$(s \in U)$ Remove$(U, s)$;
08.   Insert$(U, s, \text{Key}(s))$;
09. else if$(g(s) = rhs(s)$ AND $s \in U)$ Remove$(U, s)$;

**UpdateStateLower(s)**
10. if$(g(s) > rhs(s))$
11.   if$(s \in U)$ Remove$(U, s)$;
12.   Insert$(U, s, \text{Key}(s))$;
13. else if$(g(s) = rhs(s)$ AND $s \in U)$ Remove$(U, s)$;

**ComputeState(s)**
14. if$(s \neq s_{goal})$
15.   if(it is inital planning) $rhs(s) = \min_{(s', s'') \in conbrs(s)} \text{ComputeCost}(s, s', s'', rhs)$;
16.   else $rhs(s) = \min_{(s', s'') \in conbrs(s)} \text{ComputeCost}(s, s', s'', g)$;

**ComputeShortestPath()**
17. while$(\min_{s \in U}(\text{Key}(s)) < \text{Key}(s_{start})$ OR $rhs(s_{start}) \neq g(s_{start}))$
18.   $s = U.\text{Top}()$;
19.   if$(g(s) > rhs(s))$
20.     $g(s) = rhs(s)$;
21.     Remove$(U, s)$;
22.     for all $s' \in nbrs(s)$
23.       ComputeState$(s')$; UpdateStateLower$(s')$;
24.   else
25.     $g(s) = \infty$;
26.     for all $s' \in nbrs(s) \cup \{s\}$
27.       ComputeState$(s')$; UpdateState$(s')$;

**FindRaiseStatesOnPath()**
28. $raise = false$; $s = s_{start}$; $ctr = 0$; $loop = false$;
29. while$(s \neq s_{goal}$ AND $loop = false$ AND $ctr < maxsteps)$
30.   $x = \arg\min_{s' \in conc(s)}(c(s, s') + g(s'))$;
31.   for all $s' \in vicinity(s_x)$
32.     if($s'$ is locally inconsistent AND $s'$ has never been added
        into $U$ with local underconsistency during this replanning episode)
33.       ComputeState$(s')$; UpdateState$(s')$; $raise = true$;
34.   if($s$ is visited before) $loop = true$;
35.   $s = x$; $ctr++$;

**Main()**
36. Initialize();
37. ComputeShortestPath();
38. Path extraction;
39. Post-Smoothing;
40. forever
41.   Wait for changes in cell costs;
42.   for all cells $x$ with changed costs
43.     for each state $s$ on a corner of $x$
44.       ComputeState$(s)$; UpdateStateLower$(s)$;
45.   ComputeState$(s_{start})$; UpdateState$(s_{start})$;
46.   ComputeShortestPath(); FindRaiseStatesOnPath();.
47.   while$(raise)$
48.     ComputeShortestPath(); FindRaiseStatesOnPath();.
49.   Post-Smoothing;

**Figure 4**: The update-reducing Field D* algorithm.

path between two cell corners, it checks whether the costs of all grid cells that the original path is through are all same. If they are, the original path is kept.

## EXPERIMENTAL RESULTS

We compared the performance of URFD*, Field D* and Delayed D*. 800 different 500 × 500 random grid environments were generated: 400 environments with uniform cost grids and 400 with non-uniform cost grids. For uniform cost grid environments, four different initial percentages of obstacle cells were selected: 10%, 20%, 30% and 40%. For non-uniform cost grid environments, we assigned each traversable cell an integer cost between 1 (free space) and 15, and four different initial percentages of free space cells were selected: 90%, 70%, 50% and 30%, while the rest of the cells each got a cost (infinity or an integer between 2 and 15) randomly with the same probability (namely 1/15). For each environment, the initial task was to plan a path from the lower left corner to a randomly selected goal on the right edge. After that, we altered the costs of cells close to the agent with probability 0.1 (1.6% of the cells in the environment were changed) and had each approach repair its solution path. We use the weighted heuristic, which is described in the previous section, for URFD* and Field D*.

We selected the results in three kinds of environments (A: uniform cost grids with 10% obstacle cells. B: uniform cost grids with 30% obstacle cells. C: non-uniform cost grids with 50% free space cells) and showed them in Table 1. Four performance measures were used here: the path cost, the total number of rhs-value updates (that is, updates of the rhs-values), the total number of vertex expansions (updates of the g-values) and the runtime. Each value is a ratio of a performance measure of URFD* (or Field D*) to that of Delayed D* averaged over initial planning (or replanning) episodes. Note that in environments with more free space the runtimes of initial planning and replanning of Field D* drastically increased while those of URFD* increased much more stably. The performance in environments C shows the possibility

**Table 1:** Performance comparison among URFD*, Field D* and Delayed D* in three kinds of environments

| Initial | | Path Cost | | Rhs-values Updates | | Vertex Expansions | | Runtime | |
|---|---|---|---|---|---|---|---|---|---|
| | | Replan | Initial | Replan | Initial | Replan | Initial | Replan | Initial |
| A | Field D* | 0.9538 | 0.9544 | 6.1560 | 64.1851 | 5.2317 | 32.3386 | 5.7784 | 58.7625 |
| | URFD* | 0.9510 | 0.9519 | 2.5438 | 28.5094 | 2.2528 | 14.7824 | 2.3987 | 21.0833 |
| B | Field D* | 0.9619 | 0.9618 | 2.4143 | 10.7327 | 1.8484 | 8.9259 | 2.1848 | 13.0324 |
| | URFD* | 0.9568 | 0.9567 | 1.4702 | 4.9103 | 1.4780 | 4.2553 | 1.8199 | 5.1806 |
| C | Field D* | 0.9639 | 0.9643 | 1.3155 | 1.2895 | 1.0600 | 0.9707 | 1.2174 | 4.9793 |
| | URFD* | 0.9592 | 0.9596 | 0.8337 | 1.4234 | 0.9218 | 1.0877 | 0.7672 | 2.3122 |

that the number of updates of the rhs-values during replanning could be slightly larger than that of Field D* in some scenarios. However, since the number of vertices in the priority queue is limited by selectively processing locally underconsistent vertices, making the priority queue operations less expensive, the runtime of URFD* is still shorter than that of Field D* in those scenarios.

## CONCLUSIONS

We present URFD*, a linear interpolation-based algorithm that plans and replans any-angle paths in dynamic environments with uniform and non-uniform cost grids. It makes efforts in the reduction of updates of the rhsvalues, which contributes to the gain in efficiency. The solution paths returned by URFD* are smooth and nearoptimal. As opposed to Field D*, it performs faster planning and replanning and returns a path with lower cost and fewer heading changes. However, URFD* is not optimal either due to the linear interpolation assumption.

## REFERENCES

1.  D. Ferguson, M. Likhachev and A. Stentz, "A guide to Heuristic-Based Path Planning," Proceedings of the Workshop on Planning under Uncertainty for Autonomous Systems at the International Conference on Automated Planning and Scheduling, Monterey, 5-10 June 2005, pp. 9- 18.

2.  P. Hart, N. Nilsson and B. Raphael, "A formal basis for the Heuristic Determination of Minimum Cost Paths," IEEE Transactions on Systems Science and Cybernetics, 1968, Vol. 4, No. 2, pp. 100-107.

3.  S. Koenig, M. Likhachev and D. Furcy, "Lifelong Planning A*," Artificial Intelligence Journal, Vol. 155, No. 1-2, 2004, pp. 93-146.

4.  S. Koenig and M. Likhachev, "Improved Fast Replanning for Robot Navigation in Unknown Terrain," Proceedings of the IEEE International Conference on Robotics and Automation (ICRA 2002), Washington, 11-15 May 2002, pp. 968-975.

5.  A. Botea, M. Müller and J. Schaeffer, "Near optimal Hierarchical Path-Finding," Journal of Game Development, 2004, Vol. 1, No. 1, pp. 1-22.

6.   A. Nash, K. Daniel, S. Koenig and A. Felner, "Theta*: Any-Angle Path Planning on grids," Proceedings of the National Conference on Artificial Intelligence, 22-26 July 2007, Vancouver, pp. 1177-1183.

7.   D. Šišlák, P. Volf and M. Pěchouček, "Accelerated A* Path Planning," Springer-Verlag, Berlin, 2009.

8.   D. Ferguson and A. Stentz, "The Field D* algorithm for Improved Path Planning and replanning in uniform and non-Uniform Cost Environments," Technical Report, Carnegie Mellon University, Pittsburgh, 2005.

9.   M. W. Otte and G. Grudic, "Extracting paths from Fields Built with Linear Interpolation," IEEE/RSJ International Conference on Intelligent Robots and Systems, St. Louis, 10-15 October 2009, pp. 4406-4413.

10.  D. Ferguson and A. Stentz, "The delayed D* algorithm for Efficient Path Replanning," Proceedings of the IEEE International Conference on Robotics and Automation, Barcelona, 18-22 April 2005, pp. 2045-2050.

## CITATION

C. Zheng, J. Cai and H. Yin, "A Linear Interpolation-Based Algorithm for Path Planning and Replanning on Girds," Advances in Linear Algebra & Matrix Theory, Vol. 2 No. 2, 2012, pp. 20-24. doi: 10.4236/alamt.2012.22003.

# Non-Singularity Conditions for Two Z-Matrix Types

*Shinji Miura*
Independent, Gifu, Japan

## ABSTRACT

A real square matrix whose non-diagonal elements are non-positive is called a Z-matrix. This paper shows a necessary and sufficient condition for non-singularity of two types of Z-matrices. The first is for the Z-matrix whose row sums are all non-negative. The non-singularity condition for this matrix is that at least one positive row sum exists in any principal submatrix of the matrix. The second is for the Z-matrix A which satisfies $Ax \geq 0$ where $\exists x > 0$. Let $a_{ij}$ be the ith row and the jth column element of A, and $x_j$ be the jth element of x. Let F be a subset of $N = \{1, 2, \ldots, n\}$ which is not empty, and G be the complement of F if F is a proper subset. The non-singularity condition for this matrix is $\exists i \in F$ such that $\sum_{j \in N} a_{ij} x_j > 0$ or $\exists i \in F, \exists j \in G$ such that $a_{ij} < 0$ for $\forall F \subseteq N$. Robert Beauwens and Michael Neumann previously presented conditions similar to these conditions. In this paper, we present a different proof and show that these conditions can be also derived from theirs.

## INTRODUCTION

A real square matrix whose non-diagonal elements are non-positive is called a Z-matrix. The purpose of this paper is to show a necessary and sufficient condition for non-singularity of two types of Z-matrices.

The first is the Z-matrix whose row sums are all non-negative. In this paper, we denote this as a Non-negative Sums Z-matrix (NSZ-matrix).

The second is the Z-matrix A which satisfies $Ax \geq 0$ where $\exists x > 0$. In this paper, we denote this as a Non-negative Product Z-matrix (NPZ-matrix).

The following relation exists between these matrices.

**Theorem 1.1** An NSZ-matrix is equivalent to an NPZ-matrix where all elements of x are the same number.

**Proof.** Let $N = \{1, 2, ...., n\}$ be a set of numbers, and x be a positive vector with all elements equal to the same number $x^*$.

If $A = \left(a_{ij}\right)$ is an NSZ-matrix, the ith element of Ax is $(\sum_{j \in N} a_{ij})x^*$. $\sum_{j \in N} a_{ij} \geq 0$ for $\forall i \in N$ because A is an NSZ-matrix. Also, $x^* > 0$ from the premise. Hence, $(\sum_{j \in N} a_{ij})x^* \geq 0$ for $\forall i \in N$ is satisfied.

Therefore, A is an NPZ-matrix.

Conversely, consider that $A = (a_{ij})$ is an NPZ-matrix which satisfies $(\sum_{j \in N} a_{ij})x^* \geq 0$ for $\forall i \in N$ where $\exists x^* > 0$. If we divide both sides of $(\sum_{j \in N} a_{ij})x^* \geq 0$ by $x^* > 0$, we obtain $\sum_{j \in N} a_{ij} \geq 0$ for $\forall i \in N$. Thus, A is an NSZ-matrix. [Q. E. D.]

As Theorem 1.1 shows, the NSZ-matrix is a type of the NPZ-matrix. Therefore, if we can find a necessary and sufficient condition for non-singularity of the NPZ-matrix, we find the (a) necessary and sufficient condition for non-singularity of the NSZ-matrix automatically. However, we will prove the latter condition first, and then address the former condition.

We first state the basic propositions of linear algebra used in this paper.

The determinant of a square matrix A is denoted det A in this paper.

**Theorem 1.2** Let $A = (a_{ij})$ be a Z-matrix. Take a real number $\rho$ which is equal to or more than all diagonal elements and construct the matrix $B = \rho E - A$ where E refers to the unit matrix. B is a non-negative matrix.

**Proof.** The non-diagonal elements of B=(bij) are bij=-aij. As A is a Z-matrix, bij=-aij≥0 for $\forall i \neq j$. On the other hand, the diagonal elements of B are $b_{ii} = \rho - a_{ii}$. Since $\rho \geq a_{ii}$ for $\forall i \in N$ by the premise, $b_{ii} \geq 0$ for $\forall i \in N$. Therefore, $\forall b_{ij} \geq 0$. [Q. E. D.]

**Theorem 1.3** A non-negative square matrix B always has a non-negative eigenvalue. Let $\lambda(B)$ be the maximum non-negative eigenvalue of **B**. Then there exists a non-negative eigenvector corresponding to $\lambda(B)$.

If a Z-matrix $A = \rho E - B$ satisfies $\lambda(B) \leq \rho$, A is called an M-matrix[1].

**Theorem 1.4** An M-matrix A is non-singular if and only if $\lambda(B) < \rho$. In this case, $\rho > 0$ and all elements of the inverse of an M-matrix are non-negative. In particular, all diagonal elements of the inverse are equal to or more than $1/\rho^2$.

**Proof.** As Theorems 1.3 and 1.4 are well known, we entrust the proof to another work[3]. However, as it receives less attention that diagonal elements of $A^{-1}$ are $1/\rho$ or more, we confirm this aspect.

Regarding the inverse of an M-matrix $A = \rho E - B$, $A^{-1} = E/\rho + \sum_{k=1}^{\infty}(B^k / \rho^{k+1})$ is satisfied[4]. Thus, if we set $A=(a_{ij})$ and $B^k = \{b_{ij}^{(k)}\}$, $a_{ii} = 1/\rho + \sum_{k=1}^{\infty}\{b_{ii}^{(k)} / \rho^{k+1}\}$ is satisfied. As B is a non-negative matrix and $\rho > 0$, $b_{ii}^k / \rho^{k+1} \geq 0$ for $\forall k \geq 1$. Therefore, we can obtain

$a_{ii} = 1/\rho + \sum_{k=1}^{\infty}\{b_{ii}^{(k)} / \rho^{k+1}\} \geq 1/\rho$. [Q. E. D.]

**Theorem 1.5** If the row sums of a square matrix A are all zeroes,det A=0.

**Proof.** Let $k_j$ be the jth column vector of A. We construct the linear combination $\lambda_1 k_1 + \lambda_2 k_2 + .... + \lambda_n k_n$. If the row sums of A are all zeroes, $\lambda_1 k_1 + \lambda_2 k_2 + .... + \lambda_n k_n = 0$ when $\lambda_1 = +\lambda_2 = .... = \lambda_n = 1$. Therefore, $k_1$, $k_2$, ....., $k_n$ are linearly dependent. The determinant of a matrix whose column vectors are linearly dependent is zero[5]. [Q. E. D.]

If $B = P^{-1}AP$ holds for the square matrices A and B and a non-singular matrix P, A and B are called similar to each other.

**Theorem 1.6** When two matrices are similar, if one matrix is non-singular, the other is also non-singular.

***Theorem 1.7*** Similar matrices have the same eigenvalue[6].

Here, we define the notation for submatrices in this paper.

Let $N=\{1, 2, ...., n\}$ be a set of number of rows and columns of a square matrix A and let F be a subset of N which is not empty. $A_{FF}$ refers to a submatrix of A whose row and column elements belong to F. When F is a proper subset, we define G as the complement of F. $A_{FG}$ refers to a submatrix of A whose row elements belong to F and column elements belong to G. Similarly, $A_{FG}$ refers to a submatrix of A whose row elements belong to G and column elements belong to G, and $A_{GG}$ refers to a submatrix of A whose row and column elements belong to G. Clearly, $A_{FF}$ and $A_{GG}$ are principal submatrices. $A_{NN}$ is A itself.

Based on the above, we confirm the following basic proposition.

***Theorem 1.8*** If $A_{FG}$ is a zero matrix, $\det A = (\det A_{FF})(\det A_{GG})$.

## A NECESSARY AND SUFFICIENT CONDITION FOR NON-SINGULARITY OF THE Z-MATRIX WHOSE ROW SUMS ARE ALL NON-NEGATIVE

In this section, we discuss the non-singularity of the NSZ-matrix. We reconfirm that the NSZ-matrix is defined as the Z-matrix whose row sums are all non-negative. $A=(a_{ij})$ denotes an NSZ-matrix in this section.

***Theorem 2.1*** An NSZ-matrix is an M-matrix[8].

***Proof.*** Take an NSZ-matrix A and a real number $\rho$ which is equal to or more than all diagonal element, and construct the matrix $B = \rho E - A$. B is a non-negative matrix from Theorem 1.2. Let $b_{ij}$ be the ith row and the jth column element of B and let $x_j$ be the jth element of a non-negative eigenvector of B corresponding to $\lambda(B)$. Moreover, let $x_k$ be a maximum of $x_j$. Incidentally, multiple candidates of k may exist. In that case, one can choose any k of these. From the definition of eigenvalue and eigenvector,

$\lambda(B)x_k = \sum_{j \in N} b_{kj}x_j$ is satisfied. On the other hand, $x_j \leq x_k$ for $\forall j \in N$ because $x_k$ is a maximum of $x_j$. $\forall b_{ij} \geq 0$ because B is a non-negative matrix. From these conditions, $b_{kj}x_j \leq b_{kj}x_k$ is satisfied generally. Thus

$$\lambda(B)x_k = \sum_{j \in N} b_{kj}x_j \leq (\sum_{j \in N} b_{kj})x_k .$$

Here, we confirm $x_k > 0$. Note that $x_j \geq 0$ because it is an element of non-negative eigenvector. Therefore, if $x_k$ is zero, $\forall x_j = 0$ because $x_k$ is the maximum of $x_j$. However, the eigenvector is not a zero vector from its definition. Hence, $x_k > 0$.

Based on the above, we divide both sides of the formula $\lambda(B)x_k \leq (\sum_{j \in N} b_{kj})x_k$ by $x_k > 0$. Then, we can derive $\lambda(B) \leq \sum_{j \in N} b_{kj}$. Note that $\rho - \sum_{j \in N} b_{ij} \geq 0$ for $\forall i \in N$ because $A = \rho E - B$ is an NSZ-matrix. Then, $\lambda(B) \leq \sum_{j \in N} b_{kj} \leq \rho$. It is obvious that A satisfies the definition of an M-matrix. [Q. E. D.]

**Theorem 2.2** An NSZ-matrix is non-singular if and only if it is a non-singular M-matrix.

**Proof.** It is obvious that an NSZ-matrix is non-singular if it is a non-singular M-matrix. Conversely, if an NSZ-matrix is non-singular, it is a non-singular M-matrix by Theorem 2.1. [Q. E. D.]

Considering Theorem 2.2, we see that finding a necessary and sufficient condition for non-singularity of the NSZ-matrix equates to finding a condition that it is a non-singular M-matrix. We will show this.

**Theorem 2.3** All row sums of any principal submatrix of an NSZ-matrix are non-negative.

**Proof.** Regarding A itself, this is obvious because of the definition of
the NSZ-matrix. In the following, we show a proof for principal submatrices which are not A itself.

Let F be a proper subset of N and G be the complement of G. As
all row sums of A are non-negative, $\sum_{j \in N} a_{ij} = \sum_{j \in F} a_{ij} + \sum_{j \in G} a_{ij} \geq 0$
holds for $\forall i \in F$. Hence, $\sum_{j \in F} a_{ij} \geq -\sum_{j \in G} a_{ij}$. Because G is the complement of F, it is clear that $a_{ij}$ for $\forall i \in F, \forall j \in G$ are non-diagonal elements of the Z-matrix A. Therefore, all of these are non-positive. Then,
$\sum_{j \in F} a_{ij} \geq -\sum_{j \in G} a_{ij} \geq 0$ for $\forall i \in F$ is true. [Q. E. D.]

**Theorem 2.4** If there exists at least one principal submatrix of an NSZ-
matrix whose row sums are all zeroes, the matrix is singular.

**Proof.** If all row sums of A itself, which is one of the principal submatrices of an NSZ-matrix, are zeroes, the proposition is derived from Theorem 1.5 immediately. In the following, we show a proof for principal
submatrices which are not A itself.

Choose $F \neq \phi$ which is a proper subset of F, where the row sums of $A_{FF}$
are all zeroes.G, which is the complement of F, is also not empty.

Based on this, we will confirm that $A_{FG}$ is a zero matrix. From the definition of an NSZ-matrix,

$$\sum_{j\in N} a_{ij} = \sum_{j\in F} a_{ij} + \sum_{j\in G} a_{ij} \geq 0 \quad \text{holds} \quad \text{for } \forall i \in F. \quad \text{Then,}$$

$\sum_{j\in G} a_{ij} \geq 0$ for $\forall i \in F$ are obtained because $\sum_{j\in N} a_{ij} = 0$ for $\forall i \in F$ as defined. However, $a_{ij} \leq 0$ for $\forall i \in F$, $\forall j \in G$ because A is a Z-matrix. For these two propositions to be compatible, $a_{ij}=0$ must hold for $\forall i \in F$, $\forall j \in G$. Therefore, is a zero matrix.

Then, $\det A = (\det A_{FF})(\det A_{GG})\,||$ holds by Theorem 1.8. Because all row sums of $A_{FF}$ are zeroes, $\det A_{FF} = 0|$ |by Theorem 1.5. Thus, $\det A = 0|$.
[Q. E. D.]

**Theorem 2.5** If an NSZ-matrix is non-singular, at least one positive row sum exists in any principal submatrix of the matrix.

**Proof.** Due to the contraposition of Theorem 2.4, if an NSZ-matrix is non-singular, there does not exist a principal submatrix whose row sums are all zeroes. Then, by Theorem 2.3, at least one positive row sum exists in any principal submatrix of the matrix. [Q. E. D.]

As a result of Theorem 2.5, a necessary condition for non-singularity of the NSZ-matrix is shown. We now prove this is also a sufficient condition.

**Theorem 2.6** If at least one positive row sum exists in any principal submatrix of an NSZ-matrix, the matrix is a non-singular M-matrix.

For the proof of Theorem 2.6, we have to use inference. In the following section, we will set A as an NSZ-matrix which has at least one positive row sum in any principal submatrix. Moreover, we take a real number $\rho$ which is equal to or more than all diagonal elements and construct the matrix $A = \rho E - B$. Note that, from Theorem 1.2, B is a non-negative matrix. We now prove the following Lemmas.

**Lemma 2.7** Any row sum of all principal submatrices of B is equal to or less than $\rho$.

Proof. It is obvious that $\sum_{j \in N} b_{ij} = \rho - \sum_{j \in F} a_{ij}$ for $\forall i \in F$ from the definition of B. $\sum_{j \in F} a_{ij} \geq 0$ for $\forall i \in F$ by Theorem 2.3. Thus, $\rho - \sum_{j \in F} a_{ij} \leq \rho$. Therefore, $\sum_{j \in F} b_{ij} \leq \rho$ for $\forall i \in F$. [Q. E. D.]

**Lemma 2.8** Any principal submatrix of B has at least one row sum which is less than $\rho$.

**Proof.** If we take $\forall F \subseteq N$, $\exists i \in F$ such that $\sum_{j \in F} a_{ij} > 0$ from the premise. Hence, $\exists i \in F$ such that $\rho - \sum_{j \in F} a_{ij} < \rho$. Note that $\sum_{j \in F} b_{ij} = \rho - \sum_{j \in F} a_{ij}$ is true from the definition of B. Thus, $\exists i \in F$ such that $\sum_{j \in F} b_{ij} < \rho$ for $\forall F \subseteq N$. [Q. E. D.]

Then, we classify elements belonging to the number set N.

According to Lemma 2.8, B itself, which is one of the principal submatrices of B, has at least one row sum which is less than $\rho$. We choose $\forall i \in N$ which satisfy $\sum_{j \in G} b_{ij} < \rho$ to belong to the set. $H_1 H_2 \neq \phi$ from the premise.

If $\forall i \in N$ belong to $H_1$, the classification is complete. In the following, we consider the case where $\forall i \in N$ which does not belong to $H_1$. First, we prove the following Lemma.

**Lemma 2.9** $\sum_{j \in N} b_{ij} = \rho$ for $\forall i \notin H_1$.

**Proof.** By Lemma 2.7, $\sum_{j \in N} b_{ij} \leq \rho$ for $\forall i \in N$. Since $i \notin H_1$, $\sum_{j \in N} b_{ij} < \rho$ is not true from the definition of $H_1$. Thus, $\sum_{j \in N} b_{ij} = \rho$ for $\forall i \notin H_1$. [Q. E. D.]

We now consider the classification where $\forall i \in N$ that does not belong to $H_1$. We define F as the complement of $H_1$. By Lemma 2.8, $\exists i \in F$ such that $\sum_{j \in F} b_{ij} < \rho$. We classify such i as belonging to the set $H_1$.

If $\forall i \in N$ belong to $H_1$ or $H_2$, the classification is complete. If $\forall i \in N$ which belongs to neither set, we execute the third classification.

Generally, r classification steps are executed when $\forall i \in N$ which does not belong to any of $H_1, ..., H_{r-1}$.

In such a case, we define F as the complement of $U_{s=1}^{r-1} H_s$. Then, $\exists i \in F$ such that $\sum_{j \in N} b_{ij} < \rho$ by Lemma 2.8. We classify such i as belonging to the set $H_r$. The next Lemma is obvious from the past consideration.

**Lemma 2.10** $H_r \neq \phi$ if $\forall i \in N$ that belongs to the complement of $U_{s=1}^{r-1} H_s$.

Then, the following Lemmas are derived.

**Lemma 2.11** If $H_r \neq \phi$ for $r \geq 2$, then $H_s \neq \phi$ for $1 \leq s \leq r-1$.

**Proof.** It is obvious by the method to construct $H_r$ and mathematical induction. [Q. E. D.]

**Lemma 2.12** If $r \neq s$, then $H_r I H_s = \phi$.

**Proof.** Without loss of generality, we suppose $r>s$ and prove the Lemma under this supposition. If $i \in H_r$, $i$ belongs to the complement of $U_{s=1}^{r-1} H_s$ by the definition of $H_r$. Hence, $i$ does not belong to $H_s$ where $r>s$. That is, $H_r I H_s = \phi$. [Q. E. D.]

From Lemmas 2.11 and 2.12, $H_r$ can be defined at most by $r=n$. In short, the classification is finished in limited time. If it is finished within $m$ times, we derive the next Lemmas.

**Lemma 2.13** $N = U_{r=1}^{m} H_r$.

**Proof.** It is obvious that $N = U_{r=1}^{m} H_r$. Hence, we prove $N \subseteq U_{r=1}^{m} H_r$. We suppose that $\forall i \in N$ that belongs to the complement of $U_{r=1}^{m} H_r$. Then, $\exists i \in H_{m+1}$ from Lemma 2.10, but this contradicts the definition of $m$. By reductio ad absurdum, $\forall i \in N$ belong to $U_{r=1}^{m} H_r$. [Q. E. D.]

**Lemma 2.14** We define $G = U_{s=1}^{r-1} H_s$. If $H_r$ where $r \geq 2$ is not empty, $\sum_{j \in G} b_{ij} > 0$ for $\forall i \in H_r$.

**Proof.** Let $F$ be the complement of $G = U_{s=1}^{r-1} H_s$ and $i$ be any element of $H_r$ where $r \geq 2$. By Lemma 2.9, $\sum_{j \in N} b_{ij} = \sum_{j \in F} b_{ij} + \sum_{j \in G} b_{ij} = \rho$ holds. Then, $\sum_{j \in F} b_{ij} = \rho - \sum_{j \in G} b_{ij}$ is satisfied. $\sum_{j \in F} b_{ij} < \rho$ Holds by the definition of $H_r$. Therefore, $\rho - \sum_{j \in G} b_{ij} < \rho$ holds. Thus, we obtain $\sum_{j \in G} b_{ij} > 0$. [Q. E. D.]

Note that because B is a non-negative matrix, a non-negative eigen-vector corresponding to $\lambda(B)$ exists by Theorem 1.3. However, $\lambda(B)$ refers to a maximum non-negative eigenvalue of B. Let $x_j$ be the jth element of the non-negative eigenvector and $x_k$ be a maximum of $x_j$.

**Lemma 2.15** If $k \in H_1 \lambda(B) < \rho$.

**Proof.** $x_j \leq x_k$ for $\forall j \in N$ from the definition of k. Further, $\forall b_{kj} \geq 0$ because B is a non-negative matrix. From these two conditions, $b_{kj}x_j \leq b_{kj}x_k$ for $\forall j \in N$.

On the other hand, $\lambda(B)x_k \sum_{j \in N} b_{kj}x_j$ is true from the definition of ei-genvalue and eigenvector. From these, $\lambda(B)x_k \leq (\sum_{j \in N} b_{kj})x_k$ is derived.

Note that if $x_k$, which is a maximum of the non-negative eigenvector, is zero, the eigenvector must be a zero vector. However, this contra-dicts the definition of eigenvector. Thus, $x_k > 0$. Then if we divide both sides of the former formula by $x_k$, we derive $\lambda(B) \leq \sum_{j \in N} b_{kj}$ . Note that $\sum_{j \in N} b_{kj} < \rho$ from the premise $k \in H_1$.

From these two formulas, $\lambda(B) < \rho$ is derived. [Q. E. D.]

**Lemma 2.16** Let r be a natural number equal to or more than 2. If $k \notin U_{s=1}^{r-1} H_s$ and $k \in H_r$, then $\lambda(B) < \rho$.

**Proof.** We define $G = U_{s=1}^{r-1} H_s$ and F as the complement of G. $F \neq \phi$ since $k \in H_r$, and

$G = U_{s=1}^{r-1} H_s \neq \phi$ by Lemma 2.11. Let $h \in G$ such that $x_j \leq x_h$ for $\forall j \in G$. As $x_j$ is an element of the eigenvector corresponding to eigenvalue $\lambda(B)$, $\lambda(B)x_k = \sum_{j \in F} b_{kj}x_j + \sum_{j \in G} b_{kj}x_j$ holds. $x_j \leq x_h$ for $\forall j \in G$ from the definition of h and $b_{kj} \geq 0$ because B is a non-negative matrix. Hence, $b_{kj}x_j \leq b_{kj}x_h$ for $\forall j \in G$ holds. Therefore, $\sum_{j \in G} b_{kj}x_j \leq (\sum_{j \in G} b_{kj})x_h$ is satisfied. Further, $x_j \leq x_k$ by the definition of k and $b_{kj} \geq 0$. Accordingly, $b_{kj}x_j \leq b_{kj}x_h$ for $\forall j \in F$ holds. Hence, $\sum_{j \in F} b_{kj}x_j \leq (\sum_{j \in F} b_{kj})x_k$ is satisfied. From the above results, we see that $\lambda(B)x_k = \sum_{j \in F} b_{kj}x_j + \sum_{j \in G} b_{kj}x_j \leq (\sum_{j \in F} b_{kj})x_k + (\sum_{j \in G} b_{kj})x_h$.

We divide the leftmost and rightmost sides of this formula by $x_k > 0$, $\lambda(B)x_k \leq \sum_{j \in F} b_{kj} + (\sum_{j \in G} b_{kj})(x_h / x_k)$ is derived. Note that $x_k$ is defined as the maximum of the elements in the non-negative eigenvector corresponding to $\lambda(B)$. Moreover, k does not belong to G and h belongs to G. Hence, $x_h < x_k$. Therefore, $x_h / x_k < 1$. From the premise $k \in H_r$ where $r \geq 2$ and Lemma 2.14, $\sum_{j \in G} b_{kj} > 0$. Accordingly,

$(\sum_{j \in G} b_{kj})(x_h / x_k) < \sum_{j \in G} b_{kj}$ holds. Hence, $\lambda(B) \leq (\sum_{j \in F} b_{kj}) + (\sum_{j \in G} b_{kj})$ $(x_h / x_k) < \sum_{j \in F} b_{kj} + \sum_{j \in G} b_{kj}$ holds. As $\sum_{j \in N} b_{kj} = \sum_{j \in F} b_{kj} + \sum_{j \in G} b_{kj}$, we derive $\lambda(B) < \sum_{j \in N} b_{kj}$. Note that as $k \notin U_{s=1}^{r-1} H_s$ by the premise, it does not belong to $H_1$ either. Therefore, $\sum_{j \in N} b_{kj} = \rho$ by Lemma 2.9. $\lambda(B) < \sum_{j \in N} b_{kj} = \rho$ is derived. [Q. E. D.]

Proof of Theorem 2.6. By Lemma 2.13, $\forall k \in N$ belong to any $H_r$. By Lemmas 2.15 and 2.16, $\lambda(B) < \rho$ is true when k belongs to any $H_r$. From Theorem 1.4, A is a non-singular M-matrix. [Q. E. D.]

Now, we can show a necessary and sufficient condition for non-singularity of the NSZ-matrix. We will also show a necessary and sufficient condition for singularity of the matrix.

**Theorem 2.17** A necessary and sufficient condition for non-singularity of the Z-matrix whose row sums are all non-negative is that at least one positive row sum exists in any principal submatrix of the matrix.

**Proof.** Necessity is shown in Theorem 2.5. Sufficiency is derived from Theorem 2.6. [Q. E. D.]

**Theorem 2.18** A necessary and sufficient condition for singularity of the Z-matrix whose row sums are all non-negative is that there exists at least one principal submatrix of the matrix whose row sums are all zeroes.

**Proof.** By the contraposition of Theorem 2.17, a necessary and sufficient condition for singularity of the NSZ-matrix is that at least one principal submatrix of the NSZ-matrix whose row sums are all non-positive exists. From Theorem 2.3, this means that there exists at least one principal submatrix of the NSZ-matrix whose row sums are all zeroes. [Q. E. D.]

## A NECESSARY AND SUFFICIENT CONDITION FOR NON-SINGULARITY OF THE Z-MATRIX WHICH HAS A NON-NEGATIVE PRODUCT WITH A POSITIVE VECTOR

In this section, we discuss the non-singularity of the NPZ-matrix. We reconfirm that the NPZ-matrix is defined as the Z-matrix A which satisfies $A_x \geq 0$ where $\exists x > 0$. $A = (a_{ij})$ denotes an NPZ-matrix in this section.

We construct the diagonal matrix P whose i th diagonal element is the ith element of x. As the diagonal elements of P are all positive, its inverse $P^{-1}$ exists. Note that $P^{-1}$ is a diagonal matrix whose ith diagonal element is $1/x_j$.

We subsequently construct a matrix $V=(v_{ij})$ which satisfies $V = P^{-1}AP$. A and V are similar to each other by the definition of matrix similarity.

***Theorem 3.1*** $v_{ij} = a_{ij}x_j / x_i$ for $\forall i, j \in N$.

***Proof.*** As the ith row vector of $P^{-1}$ is $(0,...0,1/x_i,0,...,0)$ and the jth column vector of A is

$(a_{ij},...a_{i-1j},a_{ij},a_{i+1j}...,a_{nj})$, the (I,j)th element of $P^{-1}A$ is

$(1/x_i)a_{ij} + \sum_{k \neq 1} 0 a_{kj} = a_{ij}/x_i$. Thus, the ith row vector of $P^{-1}A$ is

$(a_{i1}/x_i,...a_{ij-1}/x_i, a_{ij+1}/x_i...,a_{in}/x_i)$. Further, the jth column vec-

tor of P is $(0,...,0,x_j,0...0)$. Thus, the (I,j) th element of $V = P^{-1}AP$ is

$v_{ij} = (a_{ij}/x_i)x_j + \sum_{h \neq j}(a_{ih}/x_i)0 = a_{ij}x_j / x_i$. [Q. E. D.]

***Theorem 3.2*** V is an NSZ-matrix.

***Proof.*** $a_{ij} \leq 0$ for $\forall i \neq j$ because A is a Z-matrix, and $x_i > 0, x_j > 0$ be-

cause x>0. Therefore, $a_{ij}x_j / x_i \leq 0$ for $\forall i \neq j$. Then, $v_{ij} \leq 0$ for $\forall i \neq j$ by Theorem 3.1. That is, V is a Z-matrix.

Moreover, $\sum_{j \in N} a_{ij}x_j \geq 0$ for $\forall i \in N$ because A is an NPZ-matrix. If we

divide both sides of this formula by x>0, we obtain $\sum_{j \in N} a_{ij}x_j / x_i \geq 0$

for $\forall i \in N$. From Theorem 3.1, $\sum_{j \in N} v_{ij} \geq 0$ for $\forall i \in N$ is derived.

We have shown that V is a Z-matrix and all row sums of V are non-negative. Thus, Vsatisfies the definition of an NSZ-matrix. [Q. E. D.]

**Theorem 3.3** An NPZ-matrix A is non-singular if and only if at least one positive row sum exists in any principal submatrix of V.

**Proof.** Vis an NSZ-matrix from Theorem 3.2. Then, by Theorem 2.17, V is non-singular if and only if at least one positive row sum exists in any principal submatrix of V. Since A and V are similar to each other, this is also a necessary and sufficient condition for the non-singularity of A by Theorem 1.6. [Q. E. D.]

Further, we will prove that this is also an equivalent condition that A is a non-singular M-matrix.

**Theorem 3.4** Take a real number $\rho$ which is equal to or more than all diagonal elements of A and construct the matrix $B = \rho E - A$. Moreover, we construct $W = \rho E - V$. Then, B and W are similar to each other.

**Proof.** $P^{-1}BP = P^{-1}(\rho E - A)P = \rho P^{-1}EP - P^{-1}AP = \rho E - V = W$. [Q. E. D.]

**Theorem 3.5** $W = \rho E - V$ is a non-negative matrix.

**Proof.** $\forall a_{ij} \leq \rho$ by the definition of $\rho$ and $v_{ii} = a_{ii}x_i / x_i$ from Theorem 3.1. Then, $\forall v_{ii} \leq \rho$. Therefore, $\rho - v_{ii}$, which are diagonal elements of $W = \rho E - V$, are non-negative for $\forall i \in N$. Then $-v_{ij}$, which are non-diagonal elements of $W = \rho E - V$, are non-negative for $\forall i \neq j$ because V is a Z-matrix by Theorem 3.2. Thus, W is a non-negative matrix. [Q. E. D.]

**Theorem 3.6** An NPZ-matrix is an M-matrix[9].

**Proof.** $B = \rho E - A$ and $W = \rho E - V$ are both non-negative matrices due to Theorems 1.2 and 3.5. Thus, they have maximums of non-negative eigenvalues both by Theorem 1.3. Let $\lambda(B)$ and $\lambda(W)$ each be a maximum of a non-negative eigenvalue. Because B and W are similar by Theorem 3.4, $\lambda(B) = \lambda(W)$ by Theorem 1.7. Note that V is an NSZ-matrix from Theorem 3.2. Then, V is an M-matrix from Theorem 2.1. Therefore, $\lambda(W) \leq \rho$ from the definition of an M-matrix. Then, $\lambda(B) \leq \rho$ because $\lambda(B) = \lambda(W)$. We can confirm that A satisfies the definition of an M-matrix. [Q. E. D.]

**Theorem 3.7** An NPZ-matrix is non-singular if and only if it is a non-singular M-matrix.

**Proof.** It is obvious that an NPZ-matrix is non-singular if it is a non-singular M-matrix. Conversely, if an NPZ-matrix is non-singular, it is a non-singular M-matrix by Theorem 3.6. [Q. E. D.]

By Theorem 3.3, we can find a necessary and sufficient condition for non-singularity of an NPZ-matrix. However, this condition is described with parts of V which is similar to A. It is not described with parts of A and x that are used in the definition of the NPZ-matrix. We will look for a condition described with such parts.

In the following, let F be a subset of N that is not empty, and G be the complement of F if F is a proper subset. Then, the next Theorems hold.

**Theorem 3.8** $\sum_{j \in F} v_{ij} > 0$ if and only if $\sum_{j \in F} a_{ij} x_j > 0$ for $\forall i \in F$ where $\forall F \subseteq N$.

**Proof.** This is true because $\sum_{j \in F} v_{ij} = (\sum_{j \in F} a_{ij} x_j) / x_i$ from Theorem 3.1 and $\forall x_i > 0$ because x is a positive vector. [Q. E. D.]

**Theorem 3.9** If F is a proper subset, $-a_{ij} x_j \geq 0$ for $\forall i \in F, \forall j \in G$.

**Proof.** Since G is the complement of F, $a_{ij}$ for $\forall i \in F, \forall j \in G$ are non-diagonal elements of A. Then, they are non-positive because A is a Z-matrix. Moreover, $\forall x_j > 0$ because x is a positive vector. Thus, $-a_{ij} x_j \geq 0$ for $\forall i \in F, \forall j \in G$. [Q. E. D.]

**Theorem 3.10** $\sum_{j \in F} a_{ij} x_j \geq 0$ for $\forall i \in F$ where $\forall F \subseteq N$.

**Proof.** If F=N, this is obvious because of the definition of the NPZ-matrix. In the following, we show a proof for $F \neq N$.

By the definition of an NPZ-matrix, $\sum_{j \in N} a_{ij} x_j = \sum_{j \in F} a_{ij} x_j + \sum_{j \in G} a_{ij} x_j \geq 0$ is satisfied. Hence, we obtain $\sum_{j \in F} a_{ij} x_j \geq -\sum_{j \in G} a_{ij} x_j$. Note that $-\sum_{j \in G} a_{ij} x_j \geq 0$ for $\forall i \in F$ holds from Theorem 3.9. Accordingly, $\sum_{j \in F} a_{ij} x_j \geq -\sum_{j \in G} a_{ij} x_j \geq 0$. [Q. E. D.]

**Theorem 3.11** Let i be any element of $\forall F \subseteq N$. Then, $\sum_{j \in N} a_{ij} x_j > 0$ or $\exists j \in G$ such that $a_{ij} < 0$ is a necessary and sufficient condition for $\sum_{j \in F} a_{ij} x_j > 0$.

**Proof.**

[Sufficiency] We prove this Theorem by dividing it into two cases.

1. The case $\sum_{j\in N} a_{ij}x_j > 0$.

   If F=N, this is obvious. In the following, we show a proof for $F \subset N$.

   By the supposition, $\sum_{j\in N} a_{ij}x_j = \sum_{j\in F} a_{ij}x_j + \sum_{j\in G} a_{ij}x_j > 0$ is satis-

   fied. Therefore, we obtain $\sum_{j\in F} a_{ij}x_j > -\sum_{j\in G} a_{ij}x_j$. Note that if $i \in F$,

   $-\sum_{j\in G} a_{ij}x_j \geq 0$ from Theorem 3.9. Hence, $\sum_{j\in F} a_{ij}x_j \geq -\sum_{j\in G} a_{ij}x_j \geq 0$.

2. The case $\exists j \in G$ such that $a_{ij} < 0$.

   If F=N, G cannot be defined. Thus, this case is applied only when $F \subset N$.

   Let k be one of $j \in G$ which satisfies $a_{ij} < 0$, and H be a set which removes k from G. By the definition of an NPZ-matrix,

   $\sum_{j\in N} a_{ij}x_j = \sum_{j\in F} a_{ij}x_j + \sum_{j\in H} a_{ij}x_j + a_{ik}x_k \geq 0$ is satisfied. Then we

   obtain $\sum_{j\in F} a_{ij}x_j + \sum_{j\in H} a_{ij}x_j \geq -a_{ik}x_k$. Since we consider the case $a_{ik} < 0$, and $x_k > 0$ is true from the definition of an NPZ-matrix, we obtain $-a_{ik}x_k > 0$. Then $\sum_{j\in F} a_{ij}x_j + \sum_{j\in H} a_{ij}x_j > 0$; in other words

   $\sum_{j\in F} a_{ij}x_j > -\sum_{j\in H} a_{ij}x_j$ is true. Further, considering $i \in F$ and

   $H \subset G$, $-\sum_{j\in H} a_{ij}x_j \geq 0$ is satisfied by Theorem 3.9. Therefore, we

   obtain $\sum_{j\in F} a_{ij}x_j > -\sum_{j\in H} a_{ij}x_j \geq 0$. [Q. E. D.]

[Necessity] By the definition of an NPZ-matrix, $\sum_{j\in N} a_{ij}x_j \geq 0$ for $\forall i \in F$

and $a_{ij} \leq 0$ for $\forall i \in F, \forall i \in G$ holds. Thus, the negative proposition of "

$\sum_{j\in N} a_{ij}x_j > 0$ or $\exists j \in G$ such that $a_{ij} < 0$" is "$\sum_{j\in N} a_{ij}x_j = 0$ and $a_{ij} = 0$ for

$\forall j \in G$". Further, as $i \in F$, the negative proposition of "$\sum_{j \in F} a_{ij} x_j > 0$

" is "$\sum_{j \in F} a_{ij} x_j = 0$" from Theorem 3.10. Therefore, the contraposition of this Theorem is as follows. Let $i$ be any element of $\forall F \subseteq N$. If $\sum_{j \in N} a_{ij} x_j = 0$ and $a_{ij} = 0$ for $\forall j \in G$, $\sum_{j \in F} a_{ij} x_j = 0$.

When F=N, this is obvious. When FÖN, if the supposition of the contraposition is satisfied,

$$0 = \sum_{j \in N} a_{ij} x_j + \sum_{j \in F} a_{ij} x_j + \sum_{j \in G} a_{ij} x_j = \sum_{j \in F} a_{ij} x_j + \sum_{j \in G} 0 x_j = \sum_{j \in F} a_{ij} x_j$$

is true for $\forall i \in F$ where $\forall F \subseteq N$. That is, the contraposition is true. [Q. E. D.]

Now, we can show a necessary and sufficient condition for non-singularity of the NPZ-matrix described with parts of A and x. We reconfirm that the NPZ-matrix is defined as a Z-matrix which satisfies $Ax \geq 0$ where $\exists x > 0$. Let F be a subset of the number set N which is not empty, and G be the complement of F.

**Theorem 3.12** Let the Z-matrix $A = (a_{ij})$ satisfy $Ax \geq 0$ where $\exists x > 0$. A necessary and sufficient condition for the non-singularity of A is $\exists i \in F$ such that $\sum_{j \in N} a_{ij} x_j > 0$ or $\exists i \in F, \exists i \in G$ such that $a_{ij} < 0$ for $\forall F \subseteq N$ [10].

**Proof.** By Theorem 3.3, an NPZ-matrix is non-singular if and only if $\exists i \in F$ such that $\sum_{j \in F} v_{ij} > 0$ for $\forall F \subseteq N$. By referring to Theorems 3.8 and 3.11, this condition can be rewritten as $\exists i \in F$ such that $\sum_{j \in N} a_{ij} x_j > 0$ or $\exists i \in F, \exists j \in G$ such that $a_{ij} < 0$ for $\forall F \subseteq N$. [Q. E. D.]

**Theorem 3.13** Let the Z-matrix $A=(a_{ij})$ satisfy $Ax \geq 0$ where $\exists x > 0$. A necessary and sufficient condition for the singularity of A is $\forall F \subseteq N$ such that $\sum_{j \in N} a_{ij} x_j = 0$ for $\forall j \in F$ and $a_{ij}=0$ for $\forall j \in F, \forall j \in G$ [11].

**Proof.** By the contraposition of Theorem 3.12, an NPZ-matrix is singular if and only if $\exists F \subseteq N$ such that $\sum_{j \in N} a_{ij} x_j \leq 0$ for $\forall j \in F$ and $Ax \geq 0$ for $\forall j \in F, \forall j \in G$. However, $\sum_{j \in N} a_{ij} x_j \geq 0$ for $\forall j \in F$ and

$a_{ij} x_j \leq 0$ for $\forall j \in F, \forall j \in G$ are satisfied by the definition of an NPZ-matrix. Therefore, this singularity condition means $\exists F \subseteq N$ such that $\sum_{j \in N} a_{ij} x_j = 0$ for $\forall i \in F$ and $a_{ij}=0$ for $\forall j \in F, \forall j \in G$. [Q. E. D.]

## DERIVATION FROM THE CONDITIONS BY ROBERT BEAUWENS AND MICHAEL NEUMANN

In fact, Robert Beauwens has already shown a condition which resembles what is shown in Theorem 2.17 as a necessary and sufficient condition for non-singularity of the NSZ-matrix. It is as follows.

First, we define the necessary concept.

If an n-dimensional square matrix $A=(a_{ij})$ satisfies $|a_{ii}| \geq \sum_{i \neq j} |a_{ij}|$ for $\forall i \in N$, A is called diagonally dominant.

Then, if a diagonally dominant matrix $A=(a_{ij})$ satisfies $|a_{ii}| > \sum_{j=1}^{i-1} |a_{ij}|$ for $\forall i \in N$, A is called lower semi-strictly diagonally dominant.

In the following, a permutation of A denotes $B=PAP^T$ by a permutation matrix P. Then, if B, a permutation of A, is lower semi-strictly diagonally dominant, A is called semi-strictly diagonally dominant.

Beauwens showed the following Theorem.

**Theorem 4.1** Let the Z-matrix A be diagonally dominant and have diagonal elements that are all non-negative. A is a non-singular M-matrix if and only if it is semi-strictly diagonally dominant[12].

If A is a Z-matrix, $|a_{ij}| = -a_{ij}$ holds for all non-diagonal elements. If diagonal elements of A are nonnegative, $|a_{ii}| = a_{ii}$ holds for all diagonal elements.

Thus, if A is diagonally dominant and has diagonal elements that are all non-negative, all row sums of A are non-negative. Conversely, if all row sums of the Z-matrix A are non-negative, diagonal elements of it are all non-negative by Theorem 2.3, and it is obviously diagonally dominant.

Therefore, the matrix which Theorem 4.1 addresses is nothing but the NSZ-matrix defined in this paper. Further, if $A=(a_{ij})$ satisfies the premise of Theorem 4.1, $|a_{ii}| > \sum_{j=1}^{i-1}|a_{ij}|$ can be rewritten as $\sum_{j=1}^{i} a_{ij} > 0$. Hence, Theorem 4.1 can be rewritten as follows.

**Theorem 4.2** The NSZ-matrix $A=(a_{ij})$ is a non-singular M-matrix if and only if A satisfies $\sum_{j=1}^{i} a_{ij} > 0$ for $\forall i \in N$ or $B=(b_{ij})$, a permutation of A, satisfies $\sum_{j=1}^{i} b_{ij} > 0$ for $\forall i \in N$.

Considering Theorem 2.2, Theorem 4.2 can be also rewritten as follows.

**Theorem 4.3** The NSZ-matrix $A=(a_{ij})$ is non-singular if and only if A satisfies $\sum_{j=1}^{i} a_{ij} > 0$ for $\forall i \in N$ or $B=(b_{ij})$, a permutation of A, satisfies $\sum_{j=1}^{i} b_{ij} > 0$ for $\forall i \in N$.

The Beauwens condition shown in Theorem 4.3 is equivalent to the condition shown in Theorem 2.17. However, before we prove this, we introduce another Theorem of Beauwens.

***Theorem 4.4*** The Z-matrix $A=(a_{ij})$ is a non-singular M-matrix if and only if there exists a vector $x>0$.such that $Ax \geq 0$ and $\sum_{j=1}^{i} a_{ij}x_j >0$ for $\forall i \in N$ [13].

Furthermore, Michael Neumann showed the next Theorem.

***Theorem 4.5*** Let $A=(a_{ij})$ be a Z-matrix and $B=(b_{ij})$ be a permutation of A. A is a non-singular M matrix if and only if there exists a vector $x>0$ such that $Bx \geq 0$ and $\sum_{j=1}^{i} b_{ij}x_j >0$ for $\forall i \in N$ [14].

The matrix which Theorems 4.4 and 4.5 address is nothing but the NPZ-matrix defined in this paper. Considering also Theorem 3.7, the following Theorem can be derived.

***Theorem 4.6*** The NPZ-matrix $A=(a_{ij})$ is non-singular if and only if A satisfies $\sum_{j=1}^{i} a_{ij}x_j >0$ for $\forall i \in N$ or $B=(b_{ij})$, a permutation of A, satisfies $\sum_{j=1}^{i} b_{ij}x_j >0$ for $\forall i \in N$.

The condition for non-singularity of the NPZ-matrix shown in Theorem 3.12 is equivalent to the Beauwens-Neumann condition shown in Theorem 4.6. We now prove this.

***Theorem 4.7*** Let $A=(a_{ij})$ be an NPZ-matrix, and $B=(b_{ij})$ be a permutation of i. $\exists i \in F$ such that $\sum_{j=1}^{i} a_{ij}x_j >0$ for $\forall F \subseteq N$ is a necessary and sufficient condition that $\sum_{j=1}^{i} a_{ij}x_j >0$ for $\forall i \in N$ or $\sum_{j=1}^{i} b_{ij}x_j >0$ for $\forall i \in N$.

## Proof.

[Sufficiency] Based on the premise, $\exists i \in N$ such that $\sum_{j=1}^{i} a_{ij}x_j > 0$. If we permute this i with n, $\sum_{j=1}^{i} b_{ij}x_j > 0$ is satisfied. Next, let $F_{i-1}$ be a set which removes i from N. Based on the premise, $\exists i \in F_{n-1}$ such that $\sum_{j\in F_{n-1}} a_{ij}x_j > 0$. If we permute this i with n-1, $\sum_{j=1}^{n-1} b_{ij}x_j > 0$ is satisfied. After this, in the range of $1 \leq r \leq n \leq n-2$, let $F_r$ be a set which removes i from $F_{r+1}$. Based on the premise, $\exists i \in F_r$ such that $\sum_{j\in F_r} a_{ij}x_j > 0$. If we permute this i with r, $\sum_{j=1}^{r} b_{ij}x_j > 0$ is satisfied. If these steps are executed to r=1, $\sum_{j=1}^{i} b_{ij}x_j > 0$ holds for $\forall i \in N$. [Q. E. D.]

[Necessity] $\sum_{j\in F} a_{ij}x_j \geq 0$ is guaranteed by Theorem 3.10. Then, the contraposition of the proposition is as follows. If $\exists F \subseteq N$ such that $\sum_{j\in F} a_{ij}x_j = 0$ for $\forall i \in F$, then $\exists i \in N$ such that $\sum_{j=1}^{i} a_{ij}x_j = 0$ and $\exists i \in N$ such that $\sum_{j=1}^{i} b_{ij}x_j = 0$. We prove this contraposition.

Let k be the maximum number of elements of F such that $\sum_{j\in F}^{i} a_{ij}x_j = 0$ for $\forall i \in F$. Naturally, $\sum_{j\in F}^{i} a_{kj}x_j = 0$ holds. Then, we prove $\sum_{j=1}^{i} a_{kj}x_j = 0$
.

If the number of elements of F is more than k, k cannot be the maximum number of elements of F. Hence, there is no such possibility. Thus, if we define $K = \{1, 2, L, k\}$, $F \subseteq K$ holds.

Then, we prove $\sum_{j=1}^{i} a_{kj}x_j = 0$ by dividing it into two cases.

1. The case F=K.

In this case, $\sum_{j=1}^{k} a_{kj}x_j = \sum_{j \in K} a_{kj}x_j = \sum_{j \in F} a_{kj}x_j$ is true. Moreover,

$\sum_{j \in F} a_{kj}x_j = 0$ is true from the premise. Thus, $\sum_{j=1}^{k} a_{kj}x_j = 0$.

2. The case F⊊K.

Let H be the relative complement of F in K. $\sum_{j=1}^{k} a_{kj}x_j =$

$\sum_{j \in K} a_{kj}x_j = \sum_{j \in F} a_{kj}x_j + \sum_{j \in H} a_{kj}x_j$ is true. Since $\sum_{j \in F} a_{kj}x_j = 0$ from

the premise, $\sum_{j \in K} a_{kj}x_j = \sum_{j \in H} a_{kj}x_j$. $\sum_{j \in K} a_{kj}x_j \geq 0$ because of Theo-

rem 3.10. Therefore, $\sum_{j \in H} a_{kj}x_j \geq 0$. On the other hand, $a_{kj} \leq 0$ for

$\forall j \in H$ because $k \in F$, $H \not\subset F$ and A is a Z-matrix. Further, $\forall x_j > 0$

by the definition of the NPZ-matrix. Thus, $\sum_{j \in H} a_{kj}x_j \leq 0$. In order

for these conditions to be compatible, we must have $\sum_{j \in H} a_{kj}x_j = 0$.

Therefore $\sum_{j=1}^{k} a_{kj}x_j = \sum_{j \in K} a_{kj}x_j = \sum_{j \in H} a_{kj}x_j = 0$.

Then, $\sum_{j=1}^{k} a_{kj}x_j = 0$ is proved in any case. Thus, we obtain $\exists i \in N$ such

that $\sum_{j=1}^{i} a_{ij}x_j = 0$.

Next, we consider B=(b$_{ij}$), a permutation of A. Let F′ be a permutated

set of F. Since $\exists F \subseteq N$ such that $\sum_{j \in F} a_{kj}x_j = 0$ for $\forall i \in F$ is premised

on the contraposition, then $\exists F' \subseteq N$ such that $\sum_{j \in F'} b_{ij}x_j = 0$ for $\forall i \in F'$.

Let k′ be the maximum number of elements of F′. Then, we can also

prove $\sum_{j=1}^{k'} b_{k'j}x_j = 0$ similarly to the proof for $\sum_{j=1}^{k} b_{kj}x_j = 0$. Thus, we

also obtain $\exists i \in N$ such that $\sum_{j=1}^{i} b_{ij}x_j = 0$. [Q. E. D.]

***Theorem 4.8*** Let $A=(a_{ij})$ be an NPZ-matrix, and $B=(b_{ij})$ be a permutation of A. $\exists i \in F$ such that $\sum_{j \in N} a_{ij} x_j > 0$ or $\exists i \in F, \exists j \in G$ such that $a_{ij} < 0$ for $\forall F \subseteq N$ if and only if $\sum_{j=1}^{i} a_{ij} x_j > 0$ for $\forall i \in N$ or $\sum_{j=1}^{i} b_{ij} x_j > 0$ for $\forall i \in N$.

**Proof.** This is derived from Theorems 3.11 and 4.7 immediately. [Q. E. D.]

Theorem 4.8 shows the equivalence between the two non-singularity conditions of the NPZ-matrix, the condition in Theorem 3.12 and the Beauwens-Neumann condition in Theorem 4.6. Theorem 3.12 is also derived from Theorems 4.6 and 4.8.

The equivalence between the two non-singularity conditions of the NSZ-matrix, the condition in Theorem 2.17 and the Beauwens condition in Theorem 4.3, can be also proved.

***Theorem 4.9*** Let $A=(a_{ij})$ be an NSZ-matrix, and $B=(b_{ij})$ be a permutation of A. $\exists i \in F$ such that

$\sum_{j \in F} a_{ij} > 0$ for $\forall F \subseteq N$ if and only if $\sum_{j=1}^{i} a_{ij} x_j > 0$ for $\forall i \in N$ or $\sum_{j=1}^{i} b_{ij} x_j > 0$ for $\forall i \in N$.

**Proof.** By Theorem 1.1, A is equivalent to an NPZ-matrix where all elements of x are the same number $x^*$. Therefore, considering Theorem 4.7 in the case all $x_j$ are equal to $x^*$, $\exists i \in F$ such that

$(\sum_{j \in F} a_{ij}) x^* > 0$ for $\forall F \subseteq N$ if and only if $(\sum_{j=1}^{i} a_{ij}) x^*$ for $\forall i \in N$ or $(\sum_{j=1}^{i} b_{ij}) x^* > 0$ for $\forall i \in F$. If we divide $(\sum_{j \in F} a_{ij}) x^* > 0$ and $(\sum_{j=1}^{i} a_{ij}) x^*$ and $(\sum_{j=1}^{i} b_{ij}) x^* > 0$ by $x^* > 0$, we obtain this Theorem. [Q. E. D.]

If $i \in F$, $\sum_{j \in F} a_{ij}$ means a row sum of a principal submatrix of A. Thus, $\exists i \in F$ such that $\sum_{j \in F} a_{ij} > 0$

for $\forall F \subseteq N$ means that at least one positive row sum exists in any principal submatrix of $A = (a_{ij})$. Hence, Theorem 2.17 can be also derived from Theorems 4.3 and 4.9.

## REFERENCES

1. Berman, A. and Plemmons, R.J. (1979) Nonnegative Matrices in the Mathematical Sciences. Academic Press, Cambridge.

2. Ostrowski, A. (1937-38) über die Determinanten mit überwiegender Hauptdiagonale. Commentarii Mathematici Helvetici, 10, 69-96. http://dx.doi.org/10.1007/BF01214284

3. Varga, R.S. (2000) Matrix Iterative Analysis. 2nd Revised and Expanded Edition, Springer, Berlin.

4. Nikaido, H. (1968) Convex Structures and Economic Theory. Academic Press, Cambridge.

5. DeFranza, J. and Gabliardi, D. (2009) Introduction to Linear Algebra with Applications. International Edition, The McGrow-Hill Higher Education.

6. Anton, H. and Rorres, C. (2011) Elementary Linear Algebra with Supplement Applications. International Student Version, 10th Edition, John Wiley & Sons, Boston.

7. Bretscher, O. (2009) Linear Algebra with Applications. 4th Edition, Pearson Prentice Hall, Upper Saddle River.

8.  Plemmons, R.J. (1976) M-Matrices Leading to Semiconvergent Splittings. Linear Algebra and its Applications, 15, 243-252. http://dx.doi.org/10.1016/0024-3795(76)90030-6

9.  Beauwens, R. (1976) Semistrict Diagonal Dominance. SIAM Journal on Numerical Analysis, 13, 109-112. http://dx.doi.org/10.1137/0713013

10. Plemmons, R.J. (1977) M-Matrix Characterizations. 1—Nonsingular M-Matrices. Linear Algebra and Its Applications, 18, 175-188. http://dx.doi.org/10.1016/0024-3795(77)90073-8

11. Neumann, M. (1979) A Note on Generalizations of Strict Diagonal Dominance for Real Matrices. Linear Algebra and Its Applications, 26, 3-14. http://dx.doi.org/10.1016/0024-3795(79)90168-X

# CITATION

Miura, S. (2014) Non-Singularity Conditions for Two Z-Matrix Types. Advances in Linear Algebra & Matrix Theory, 4, 109-119. doi: 10.4236/alamt.2014.42009.

# Index